Mykhailo Azarkh

In-cell approaches in electron paramagnetic resonance spectroscopy

Mykhailo Azarkh

In-cell approaches in electron paramagnetic resonance spectroscopy

In-cell approaches in electron paramagnetic resonance spectroscopy to study conformations of DNA G-quadruplexes

Südwestdeutscher Verlag für Hochschulschriften

Impressum / Imprint
Bibliografische Information der Deutschen Nationalbibliothek: Die Deutsche Nationalbibliothek verzeichnet diese Publikation in der Deutschen Nationalbibliografie; detaillierte bibliografische Daten sind im Internet über http://dnb.d-nb.de abrufbar.
Alle in diesem Buch genannten Marken und Produktnamen unterliegen warenzeichen-, marken- oder patentrechtlichem Schutz bzw. sind Warenzeichen oder eingetragene Warenzeichen der jeweiligen Inhaber. Die Wiedergabe von Marken, Produktnamen, Gebrauchsnamen, Handelsnamen, Warenbezeichnungen u.s.w. in diesem Werk berechtigt auch ohne besondere Kennzeichnung nicht zu der Annahme, dass solche Namen im Sinne der Warenzeichen- und Markenschutzgesetzgebung als frei zu betrachten wären und daher von jedermann benutzt werden dürften.

Bibliographic information published by the Deutsche Nationalbibliothek: The Deutsche Nationalbibliothek lists this publication in the Deutsche Nationalbibliografie; detailed bibliographic data are available in the Internet at http://dnb.d-nb.de.
Any brand names and product names mentioned in this book are subject to trademark, brand or patent protection and are trademarks or registered trademarks of their respective holders. The use of brand names, product names, common names, trade names, product descriptions etc. even without a particular marking in this works is in no way to be construed to mean that such names may be regarded as unrestricted in respect of trademark and brand protection legislation and could thus be used by anyone.

Coverbild / Cover image: www.ingimage.com

Verlag / Publisher:
Südwestdeutscher Verlag für Hochschulschriften
ist ein Imprint der / is a trademark of
AV Akademikerverlag GmbH & Co. KG
Heinrich-Böcking-Str. 6-8, 66121 Saarbrücken, Deutschland / Germany
Email: info@svh-verlag.de

Herstellung: siehe letzte Seite /
Printed at: see last page
ISBN: 978-3-8381-3440-6

Zugl. / Approved by: Konstanz, Universität, Diss., 2012

Copyright © 2012 AV Akademikerverlag GmbH & Co. KG
Alle Rechte vorbehalten. / All rights reserved. Saarbrücken 2012

Table of contents

Abstract…………………………………………………………..	5
Introduction………………………………………………………..……	7

I. Literature overview

1. DNA G-quadruplexes………………………………………………..	9
1.1 Structural characterization……………………...……………….	9
1.2 Human telomeric DNA……………….…………………………...	12
1.3 Methods to study G-quadruplexes….….………………………..	16
2. Spin-label EPR spectroscopy…………………………………………...	19
2.1 Resonance phenomenon………………………………………...	19
2.2 Time-domain (pulse) EPR………………………………………..	21
2.2.1 Pulses and pulse sequences…………………………….	21
2.2.2 Double electron-electron resonance…………………….	24
2.3 Nitroxides as spin labels…………………………………………...	27
2.4 EPR on spin labeled DNA……………………………………….	32
2.4.1 Site-directed spin labeling of oligonucleotides………….	32
2.4.2 DEER studies of DNA………………………………….	37
3. *In-cell* magnetic resonance………………………………………….	42
3.1 General prerequisites and techniques………………………….	42
3.2 Chemical reactions and stability of nitroxides in biological systems…………………………………………………………...	43

II. Experimental

4. Materials…………………………………………………………	49
4.1 Chemicals…………………………………………………….	49
4.2 Spin labeled DNA………………………………………….....	49
4.3 *Xenopus laevis* oocytes and cell extract……………………….	53
5. Methods………………………………………………………….	55
5.1 CD spectroscopy……………………………………………..	55
5.2 FRET………………………………………………….……..	55
5.3 Cw EPR……………………………………………………...	56
5.4 Pulse EPR: DEER and measurements of relaxation times…….	57

III. Results and Discussion

6. Human telomeric DNA in aqueous buffer solutions………………	61
6.1 Spin labeling of human telomeric DNA oligonucleotides…….	61
6.2 G-quadruplex conformations in Na^+-containing and K^+-containing solutions………………………………………..	70
6.3 Selective identification of G-quadruplexes within a long DNA sequence……………………………………………………...	76
7. *In-cell* EPR…………………………………………………….	84
7.1 Nitroxides inside oocytes an in oocyte extract of *X. laevis*……	84
7.2 *In-cell* DEER on model systems…………………………….	92
7.3 Conformations of the human telomeric repeat *in cellulo*……..	99

Conclusions………………………………………………………	111
List of references…………………………………………………	115

Additional content

List of abbreviations.. 128
Supporting material.. 130
Published results... 144

Acknowledgements... 145

Abstract

G-quadruplexes are secondary structures of nucleic acids and can be formed by DNA or RNA sequences with high content of guanines. A single-stranded overhang of the telomeric caps at human chromosomes is a guanine-rich DNA sequence consisting of d[GGGTTA] repeats and is prone to fold into G-quadruplexes. These structures are supposed to play an important role in cancer and cell cycle.

Site-directed spin labeling in combination with EPR was utilized to probe G-quadruplex topologies and their conversions upon different external factors as type of alkali ions, flanking nucleotides and neighboring quadruplexes both in buffer solutions and inside cells. In these studies mainly the human telomeric repeat d[AGGG(TTAGGG)$_3$] was investigated which is the shortest fragment of the human telomeric DNA able to form intramolecular G-quadruplexes. It was shown to adopt the antiparallel conformation in Na^+-containing solution as judged by high-resolution NMR. This sequence was reported to adopt the parallel-propeller form in the presence of K^+ (as determined from X-ray crystallography), however, the exact structure in K^+ solution remained unclear.

Different G-quadruplex assemblies were determined in EPR by measuring interspin distance distributions between two spin labels incorporated into a DNA sequence. Those distances between nitroxide-spin-labeled thymidine analogs at 5- and 11-positions in the human telomeric repeat are different for known G-quadruplexes and, as predicted from PDB data, allow for discrimination between different conformations. The measured interspin distances for a quadruplex in Na^+-containing buffer are in good agreement with expected for the antiparallel-basket structure. Analysis of distance distributions for K^+ solution revealed two peaks and lead to conclusion of 1:1 mixture of coexisting parallel and antiparallel topologies.

Spin-label EPR distance measurements were utilized to investigate a particular structure within a DNA sequence of several quadruplexes. Pairs of nitroxide spin labels were introduced in selected sites of the sequence which formed three adjacent G-quadruplexes. This allowed for observation of folding of solely terminal or middle part of this DNA. In Na^+ solution again only one type of folding – the antiparallel-basket – was observed. In contrary, neighboring quadruplexes, if observed in K^+-containing solution, appeared to play crucial role in determination of the overall folding topology. No mixture but rather formation of the single (3+1) hybrid topology was found.

EPR-based distance measurements were extended to in-cell applications. First, evaluation of nitroxides for experiments in intracellular environment of Xenopus laevis oocytes was performed. The five member ring nitroxide appeared to be much more stable than its six member ring analog. A deeper insight and analysis of the reduction kinetics was done in the oocytes cell extract and evidenced that the reduction process could be described within the frame of the Michaelis-Menten formalism and thus could be attributed to be an enzyme-mediated one.

Folding kinetics of the human telomeric repeat $d[AGGG(TTAGGG)_3]$ was monitored in cell extract by time-resolved distance measurements. Analysis of G-quadruplex conformations inside injected oocytes showed that observations in K^+-containing buffer held true for in-cell experiments as well as in-extract studies. Thus dominating role of K^+ ions in determination of overall folding of the human telomeric repeat in cellulo was established.

Introduction

All kinds of spectroscopy can be classified with respect to photonic energies involved in a typical quantum jump. Electron paramagnetic resonance (EPR) exploits electromagnetic irradiation in the microwave range (λ = 3-300 mm) is used to study energy required for reorientation of an electronic magnetic moment in a magnetic field. EPR, first developed and used by physicists, has soon found applications in a variety of fields from physics through chemistry to biological sciences and medicine.

The essential property of EPR is to detect the presence of unpaired electrons in a sample. Further, unpaired electrons can be characterized in terms of their concentration and interactions that they undergo with their local environment. Thus, being able to detect concentrations of unpaired electrons down to nanomolar level in any media like frozen and liquid solution, crystalline and amorphous material without modifying the substance in question, EPR provides a unique tool for probing paramagnetic species in all kinds of substances.

In biological systems an EPR signal can originate either from transition metal ions which are present in metal-containing enzymes and proteins or from organic radicals which, however, are often transient short-living and highly reactive species. The majority of biological samples including nucleic acids, lipids, small, large and gigantic unilamellar vesicles, membranes and proteins are diamagnetic and, as consequence, EPR-silent. To enable studying such systems by EPR paramagnetic species must and can be introduced externally as spin probes or spin labels (commonly these are nitroxides – stable organic radicals). The EPR responses from spin probes deliver information on such environmental characteristics as viscosity, pH, and polarity. In the other case, a spin label which is covalently attached to a molecule of interest enables observation of its tumbling and to gain information about overall dynamics, spatial distribution and structure of the spin labeled macromolecule.

A structure of a particular biomacromolecule is in fact that starting point where understanding of its function begins. The most desired high-resolution structure as derived from X-ray crystallography or NMR cannot be obtained in a number of cases where a biomacromolecule should be observed in a complex environment (*e.g.* in a membrane). In such situations the ability of EPR to estimate long-range distances in the nanometer range is of utmost importance and relevance. Distance constraints in EPR are commonly determined by measuring a dipole-dipole coupling between two electron spins. That is, in a biomacromolecule of interest with two spin labels at desired/characteristic sites a distance measurement provides information on a global structure of the sample, structural changes or, by means of distance distribution, the flexibility of spin labeled domains.

An ability of EPR to focus solely on paramagnetic species, remaining passive towards most aspects of the diamagnetic environment, suggests that behavior and interactions of spin labeled species can be studied in systems as complex as membranes, protein complexes, viruses and even whole cells and living organisms. From the time as early as 1965 when the first nitroxide spin probes were introduced by McConnell there were many attempts to probe intracellular environment with nitroxides, to investigate nitroxides' uptake and metabolism and to prolong life-time of nitroxides. Further, *in vivo* EPR has been introduced. This opens avenues to transfer the ability of EPR in revealing structures and characterizing structural changes on molecular level to *in-cell* samples. Consequently, experimental set up, parameters and sample preparation procedures should be evaluated and optimized allowing unraveling structures of biomacromolecules in the unperturbed intracellular environment as complex it might be.

I. Literature overview

1. DNA G-quadruplexes
1.1 Structural characterization

From the structural point of view a deoxyribonucleic acid (DNA) is a biopolymer which is built up from nucleotides (Fig. 1a). Each nucleotide, being the simplest DNA structural unit, contains three fragments: a 2′-deoxyribose moiety, a phosphodiester group and a nucleobase. There are only four types on nucleobases in DNA which, by their chemical origin, belong either to purines, namely adenine (A) and guanine (G), or to pyrimidines, namely thymine (T) and cytosine (C), (Fig. 1b-e). The nucleobases are glycosidicly attached to the 2′-deoxyribose and consequently are connected with each other via phosphodiester groups forming a sequence which represents the primary structure of DNA [1].

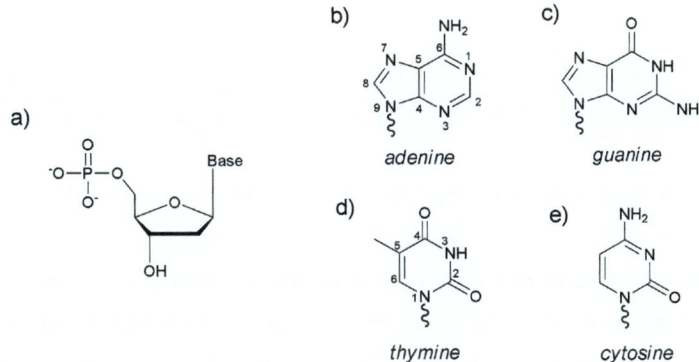

Figure 1. **Structural elements of DNA.** (a) A nucleotide and (b-e) nucleobases.

The secondary DNA structure is formed and stabilized by base-pairing – hydrogen bonds that are formed between nucleobases. According to Watson and Crick, each purine can be paired with a pyrimidine base. Pairs of A–T and G–C connected by two or three hydrogen bonds, respectively, (Fig. 2a, b) lead to DNA duplexes which, if formed within a single sequence, are represented by hairpin structures [2], by two sequences organized as a DNA double helix [1] or may yield 3D-arrangements [3] if more complementary strands are combined together. Further possibilities for connections between nucleobases are realized by Hoogsteen base-pairing when N1 and N2 amino atoms act as donors while N7 and O6 as acceptors of protons in hydrogen bonds [4].

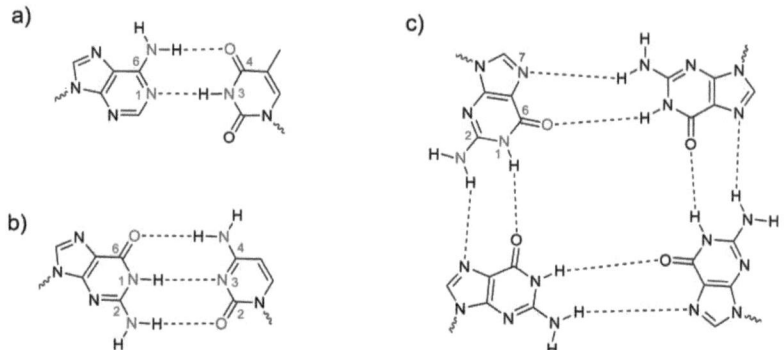

Figure 2. **Hydrogen bonding in DNA.** Watson-Crick base-pairing (a) A–T and (b) G–C. (c) G-tetrad.

Hoogsteen base-pairing allows for arrangement of four guanines in a plane unit called G-tetrad (Fig. 2c) where each of guanines is connected to the other by two hydrogen bonds. Several G-tetrads stacked upon each other and stabilized by π-π interaction form a core of a G-quadruplex structure. These structures are in particular characteristic for DNA sequences with high content of guanines and display remarkable structural polymorphism [5].

Topology of G-quadruplexes can be analyzed and classified in terms of number and orientation of DNA strands, type of loops, number of G-tetrads and orientation of the glycosidic conformation of guanines around G-tetrads.

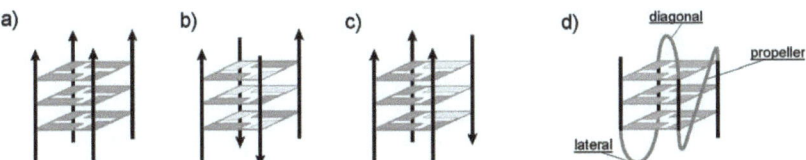

Figure 3. **Topology of G-quadruplexes.** Strand orientation: (a) parallel, (b) antiparallel and (c) (3+1) hybrid. (d) Loop types.

When all strands in a G-quadruplex have the same orientation, as counted from 5'- to 3'-end, the structure is referred to as a parallel one (Fig. 3a). Alternating orientation of the strands is characteristic for the antiparallel topology (Fig. 3b). Structures with the strand orientation that can be attributed neither to the parallel nor to the antiparallel conformation (*e.g.* three of four strands are pointing in the same direction) represent a (3+1) hybrid type (Fig. 3c). In G-quadruplexes formed of four separate strands no loops are present, but those structures where only two or even one G-rich sequence is folded several types of loops can be present (Fig. 3d). The loop which connects two G-tracts on the same side of the G-core is referred to as lateral one. Diagonal loop connects two G-tracts on the two opposite edges of the G-tetrad and runs above or below the tetrad. The propeller-type loop connects G-tracts from different G-tetrads and runs along the groove of the G-quadruplex. The loops can contain from one to many nucleosides, however, very long loops destabilize quadruplexes [6].

Different orientation of strands in a G-quadruplex force guanine bases to adopt a particular *syn-* or *anti*-conformation with respect to the sugar fragment of the nucleotide in order to match the Hoogsteen bonding within a G-tetrad (Fig. 4). Typically, in the parallel quadruplex all guanines are in *anti*-conformation.

And for example, in the antiparallel quadruplex of two DNA strands with two lateral loops and three G-tetrads the glycosidic conformations are: *anti•syn•syn•anti* in the middle; and *syn•anti•anti•syn* for the two others [5].

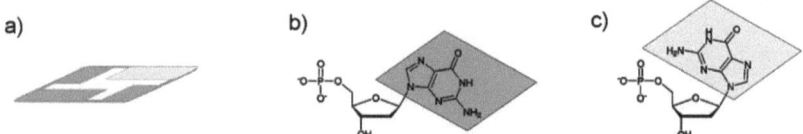

Figure 4. **Glycosidic conformations of guanines.** Schematic representation of (a) a G-tetrad, (b) *anti* and (c) *syn* guanines.

Besides **inter**molecular quadruplexes (Fig. 3a-c) which consist of four or two strands, the topological polymorphism for **intra**molecular quadruplexes is much more pronounced. A particular structure is defined by the balance of several forces which easily varies upon environment and includes metal ions present, degree of hydration, crowding effects as well as interaction with other small molecules, binding ligands or proteins [5].

1.2 Human telomeric DNA

Telomeres are DNA sequences at the ends of chromosomes (Fig. 5) [7]. Over its length, telomeric DNA is a duplex of a G-rich and a C-rich strands [8]. However, owing to a shorter length of the C-rich strand, there exists a single-stranded 3′-overhang. In humans, this overhang consists of a sequence of d[TTAGGG] hexanucleotide repeats and reaches a total length of up to 200 nucleotides [9]. The high content of guanines within this strand enables forming G-quadruplexes that are supposed to play significant roles in cell cycle control [10] and cancer [11].

A linear chromosome cannot be fully replicated at the 5'-end and, as a consequence, a telomeric DNA gets shortened upon each cell division [12]. Thus, when the telomeric DNA reaches its critical length after a certain number of divisions, the cell undergoes apoptosis or programmed cell death [13]. However, the apoptosis can be avoided if the telomeric ends are elongated after cell division. In particular, this is the function of the enzyme telomerase which is highly expressed in cancer cells [14]. The telomerase activity is essential for proliferation of cancer cells and, therefore, inhibition of this enzyme could stop tumor growth [15]. G-quadruplexes are concerned to be of anticancer activity because their formation within telomeres was shown to inhibit telomerase [16].

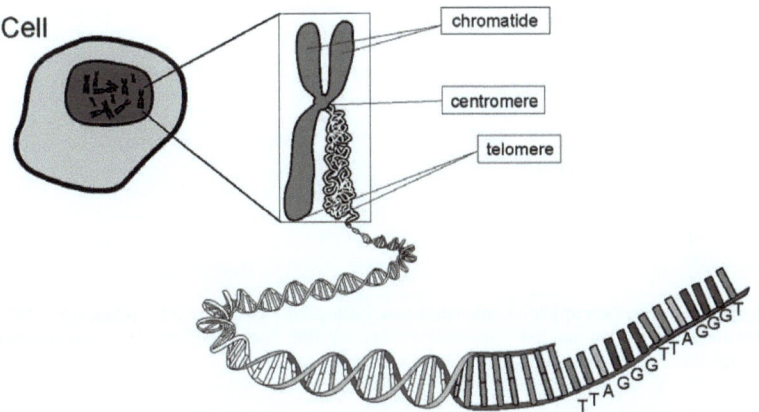

Figure 5. Human telomeric DNA. Nucleotides in the G-rich single-stranded overhang are marked with respective letters.

Calculation of the number of sequences containing four tracks of three or more guanines separated by at least one-base loops have shown that over 376 000 G-quadruplexes could be simultaneously formed in the human genome [17, 18]. Those G-quadruplexes may feature different topologies as becomes obvious from their structural polymorphism and scope of environmental factors that define a particular folding (see Section 1.1). While a certain G-quadruplex structure adopted by the human telomeric DNA *in vivo* is still not reported, the intramolecular G-quadruplexes formed by the human telomeric repeat consisting of four d[GGG] stretches were intensively studied under various conditions *in vitro*.

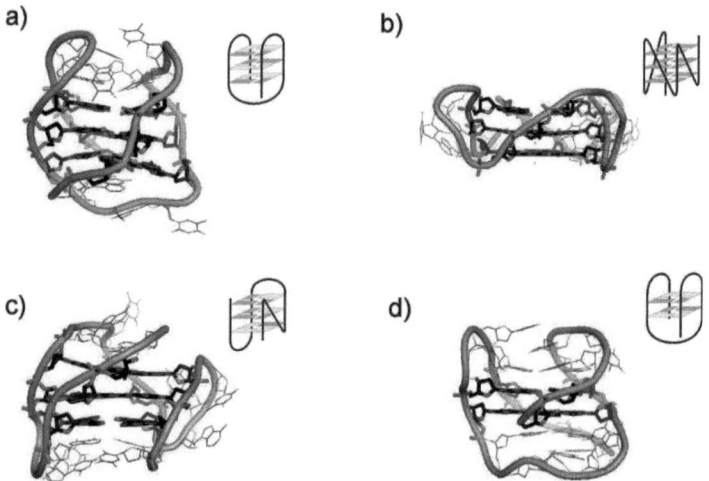

Figure 6. G-quadruplex conformations adopted by human telomeric DNA oligonucleotides. Antiparallel "basket"-type (a), parallel "propeller"-type (b), (3+1) hybrid (c), and antiparallel "basket"-type with two G-tetrads (d).

The d[AGGG(TTAGGG)$_3$] sequence in Na$^+$-containing solution was reported to fold into an antiparallel conformation (Fig. 6a, PDB code: 143D) [19]. This G-quadruplex structure was obtained high-resolution NMR and featured alternating orientation of strands and three TTA-loops in the following order: lateral – diagonal – lateral. Due to the characteristic diagonal loop this structure is also referred to as "basket"-type. The same oligonucleotide sequence was shown to adopt an absolutely different conformation in the presence of K$^+$ ions.

X-ray crystallographic studies suggested a parallel G-quadruplex which possessed all strands of the same polarity connected by three propeller-like loops (Fig. 6b, PDB code: 1KF1) [20]. The loops' type defines also the "propeller"-type conformation. Another type of intramolecular G-quadruplex was reported from an NMR study in K^+-solution on a slightly modified sequence of the human telomeric repeat. The new flanking nucleotides TT instead of A at 5'-end and an additional A at the 3'-end in the d[TTGGG(TTAGGG)$_3$A] sequence appeared to play a crucial role for the overall folding. The observed type of G-quadruplex is referred to as (3+1) hybrid structure and possesses three parallel and one antiparallel strand, one propeller and two lateral loops, one *anti•syn•syn•syn* and two *syn•anti•anti•anti* G-tetrads (Fig. 6c, PDB code: 2GKU) [21]. The fourth intramolecular G-quadruplex of the human telomeric DNA oligonucleotide was reported for the sequence d[GGG(TTAGGG)$_3$T] which has only one flanked thymidine at the 3'-end. In K^+-solution it adopts an antiparallel-basket conformation with the same strand orientation and loop types as in Na^+ structure of d[AGGG(TTAGGG)$_3$]. The characteristic feature of this fourth G-quadruplex type is that it contains only two tetrads in the G-core. The guanines which are not arranged in G-tetrads form GGG and GAA triplexes stabilized by the network of Hoogsteen hydrogen bonds and stacked at the top and the bottom of the two-tetrad G-core, respectively (Fig. 6d, PDB code: 2KF8) [22].

Despite of several G-quadruplex structures of human telomeric repeat resolved by both X-ray and NMR, the question about physiologically relevant conformation remains open. Obviously, much attention should be paid to those by K^+ induced quadruplexes. This relies on a fact of higher intracellular concentration of K^+ in comparison to Na^+ [23]. Moreover, even in the presence of both alkali ions potassium plays dominating role and determines the overall folding [23]. The propeller structure, as reported from X-ray crystallography, could be not predominant in solution but might rather appear to be that one which could be crystallized. Circular dichroism spectra of the human telomeric sequence in potassium solution display features characteristic for both parallel and antiparallel topologies. It suggests that there could be a mixture of two conformations which can even undergo interconversion as the energetic cost for it is only ~2.5 kcal/mol [24]. The (3+1) hybrid could potentially be also formed by d[AGGG(TTAGGG)$_3$] in K$^+$-containing solution but no NMR studies report on this structure yet.

The influence of flanking sequences on a G-quadruplex structure is obvious as well. Particular nucleobases at the end of the human telomeric sequence may form base pairs, triples or just stack on a G-core and, consequently, stabilize a certain conformation. Further, crowding effects may drive a topological conversion by creating conditions that favor compact conformations or associated complexes [25].

1.3 Methods to study G-quadruplexes

Two physical methods are mainly used to obtain full high-resolution structures of G-quadruplexes: X-ray crystallography and NMR. The first method is utilized for those samples which can be crystallized [26]. However, the biological relevance of the G-quadruplexes within the crystal should be concerned critically. It may appear that conditions used to grow single crystals either drive conformational interconversion or simply select a particular conformation which could be not a dominant species in solution. In contrast to X-ray crystallography, NMR facilitates structural investigation of G-quadruplexes *in vitro* [27]. In most favorable cases, when an oligonucleotide adopts a single and kinetically stable conformation in solution its high resolution structure can be obtained from NMR studies. Less favorable cases of a single kinetically unstable G-quadruplex or co-existence of several conformations limits structural analysis. Still, NMR remains suitable for observation of G-quadruplexes - without yielding detailed structural information - in such complex systems and, as was reported recently, even in intact cells [28].

Besides those two powerful high-resolution but rather time-consuming techniques there exist a number of express methods for probing G-quadruplex conformations.

Circular dichroism (CD) is utilized to distinguish between parallel and antiparallel conformations [29]. There are two strong bands at 240 nm (negative) and 260 nm (positive) characteristic for the parallel structure. The antiparallel type of folding can be recognized by strong negative band at 260 nm and positive at 295 nm. While these two conformations can be easily identified by CD spectroscopy, CD spectra of other quadruplex structures could be over-interpreted. For example, the (3+1) hybrid conformation displays a CD spectrum with features from both the parallel and the antiparallel topologies and thus is very similar to the spectrum of the mixture of these two conformations. Interpretation of CD spectra is still an empirical one. Recent theoretical treatments, however, show that different CD features may arise primarily due to different stacking orientation between adjacent G-tetrads rather than from relative orientation of the DNA strands [31].

Steady-state fluorescence is used to characterize loop regions in a quadruplex [24]. The experiment requires incorporation of the modified base 2-aminopurine instead of adenine in the oligonucleotide sequence. Its fluorescence provides very sensitive response of the environment as it is quenched upon stacking with other bases. Förster resonance energy transfer (FRET) is utilized to follow quadruplex folding [32]. The oligonucleotide under investigation is be labeled by a FRET pair: fluorophore (*e.g.* FAM) and quencher (*e.g.* TAMRA) [33].

Upon folding the distance between the fluorophore and the quencher becomes shorter than in a random coil and the FRET effect takes place: There is observed a shift of the emitted signal to higher wavenumbers. It should be noted that both a fluorophore and a quencher are rather bulky molecules and could potentially perturb the initial structure of an oligonucleotide.

A type of a G-quadruplex can be judged from ^{125}I-radioprobing experiment [34]. It is based on property of radioiodine, already introduced in the DNA structure, to induce breaks in the DNA backbone. The probability of strand breaks is inversely proportional to the distance between the radionucleotide and the sugar unit of the DNA where the break occurs. Thus, conformation of the backbone can be obtained from analysis of distributions of breaks.

Raman spectroscopy is used to probe the phosphodiester backbone conformation, the hydrogen-bonding interactions and the glycosyl conformation [35]. The antiparallel quadruplex can be identified by two characteristic bands at 1325 and 1333 cm^{-1} which arise from C2'-*endo*/*syn*-dG and C2'-*endo*/*anti*-dG, respectively.

Further, mass spectrometry [36], surface plasmon resonance (SPR) [37], polyacrylamide gel electrophoresis (PAGE) [38] and analytical centrifugation [24] can deliver information on whether DNA is folded to single or multiple G-quadruplexes and sometimes also allows for concluding about a particular type of a G-quadruplex in solution.

2. Spin-label EPR spectroscopy

2.1 Resonance phenomenon

An electron possesses an intrinsic spin angular momentum $\vec{S}$ associated with the magnetic moment $\vec{\mu}$ which is given by $\vec{\mu} = g\beta_e\vec{S}$ with the g-factor and the Bohr magneton β_e. Once the external magnetic field is applied a unique direction is defined – the direction of this magnetic field and is usually denoted as z-axis. An electron placed in such a field tends to align itself with its magnetic moment $\vec{\mu}$ along this unique direction. For $S = \frac{1}{2}$ there are two possible orientations, that is, parallel and antiparallel to the magnetic field vector. A classical magnetic moment which is not ideally aligned along the direction of an external magnetic field executes a precessional motion around it (Fig. 7a) [39-41].

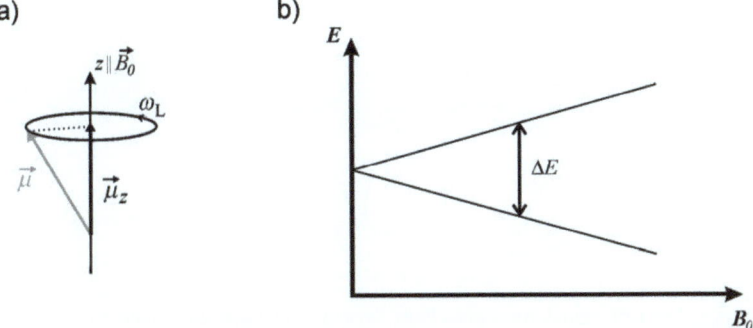

Figure 7. **Magnetic moment in the external magnetic field.** (a) Precession of a magnetic moment $\vec{\mu}$ in an external magnetic field. (b) Zeeman effect.

The two energy levels of an electron in a magnetic field, corresponding to quantum states $m_s = \frac{1}{2}$ and $m_s = -\frac{1}{2}$, are differently populated and split by the energy difference ΔE. According to the Zeeman effect, this energy difference can be driven by the strength of the applied external magnetic field (Fig. 7b). Consequently, the photon energy which is needed to induce transition between the Zeeman energy levels is studied in EPR It corresponds to irradiation that of the precession frequency $\omega_L = \frac{\Delta E}{\hbar}$ (Larmor frequency) of the magnetic moment $\vec{\mu}$ in the external magnetic field (Fig. 7a). At magnetic field values of 0.1-1.5 T the resonance frequency $v = \omega_L / 2\pi$ lies in the range of 2.8 to 42 GHz that corresponds to the microwave region.

In a continuous wave (cw) EPR experiment the net absorption of the incoming microwave irradiation from a continuous wave source is detected as a function of the applied magnetic field [39-41]. For the sake of better resolution the lock-in technique is used registering the first derivative of the absorption curve (Fig. 8).

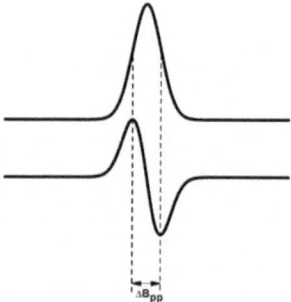

Figure 8. **The EPR signal.** Absorption line (top) and its first derivative (bottom).

Information that can be directly read out from the spectrum is number and relative intensity of EPR lines, the individual linewidth ΔB_{pp}, the amplitude (correlated with the number of EPR-active species), and the position of the line on the B_0-axis.

$$\Delta E = h\nu = g\beta B_0 \qquad (1)$$

As it is evident from eq. 1, the position of the resonance line B_0 at a given frequency ν is determined by the g-factor. Its value is never known *a priori* and depends on a particular sample. Deviations of the g-factor from the value for the free electron (g = 2.0023) are due to spin-orbit coupling. Hence, those deviations are large for electrons associated with transition metal ions and rather small (± 0.01) for organic radicals [41, 42].

2.2 Time-domain (pulse) EPR

2.2.1 Pulses and sequences

An electron in a real sample is not isolated but rather exposed to variety of interactions with its environment. As a consequence of those interactions the electron Zeeman levels are additionally split or shifted and give rise to new (often unresolved) signals in a cw EPR spectrum complicating it significantly. The energies of those levels are eigenfunctions of the spin Hamilton operator H_0 which, additionally to electron Zeeman interaction, has contributions from hyperfine, zero-field splitting, nuclear Zeeman, nuclear quadrupole and nuclear-nuclear interactions (eq.2) [42]. In general, it is not possible to extract all interactions by cw EPR.

$$\begin{aligned}H_0 &= H_{EZ} + H_{HF} + H_{ZFS} + H_{NZ} + H_{NQ} + H_{NN} = \\ &= \beta_e \tilde{\mathbf{B}}_0 \mathbf{g} \mathbf{S} + \sum_{k=1}^{m} \tilde{\mathbf{S}} \mathbf{A}_k \mathbf{I}_k + \tilde{\mathbf{S}} \mathbf{D} \mathbf{S} - \beta_n \sum_{k=1}^{m} g_{n,k} \tilde{\mathbf{B}}_0 \mathbf{I}_k + \\ &+ \sum_{I_k > 1/2} \tilde{\mathbf{I}}_k \mathbf{P}_k \mathbf{I}_k + \sum_{i \neq k} \tilde{\mathbf{I}}_i \mathbf{d}^{(i,k)} \mathbf{I}_k\end{aligned} \qquad (2)$$

Introduction of pulses in EPR allows manipulating electron spins in order to observe only desired interactions. In terms of the spin Hamiltonian formalism, unwanted parts of H_0 are removed, suppressed or averaged thus simplifying the EPR spectrum (focusing on particular observables) [43].

A pulse in time-domain EPR is a microwave irradiation of a certain power which is finite in length on the time scale.

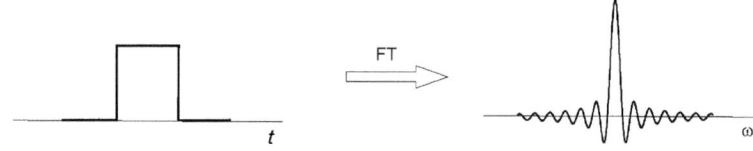

Figure 9. A rectangular pulse in time- and frequency-domains.

A pulse is characterized by an excitation width which can be obtained by Fourier transformation (FT) from the time- to a frequency-domain. A typical rectangular pulse of 32 ns corresponds to a *sinc* function with the width (half width at half height - HWHH) of 26 MHz in a frequency-domain (Fig. 9) [44]. As the majority of EPR spectra are spanned over several hundreds of MHz (*e.g.* ~200 MHz for nitroxide) EPR pulses can excite only a part of the spectrum and therefore are referred to as selective ones.

The second feature of the pulses is a flip-angle that they induce on the net magnetization vector $\vec{M}$ in a defined direction (Fig. 10). The system restores initial equilibrium state of $\vec{M}$ via relaxation pathways. This enables one to perform a pulse experiment in repetitive manner to acquire better signal at lower noise but, at the same time, limits application of long pulse sequences.

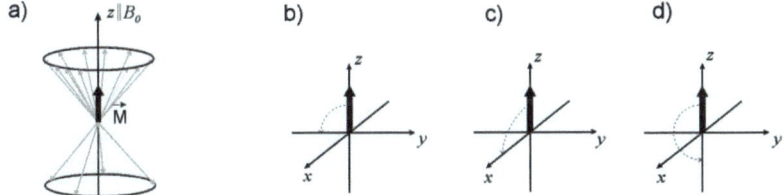

Figure 10. **Manipulation of the net magnetization vector.** Origin of the net magnetization vector (a) and its orientation after $\left(\frac{\pi}{2}\right)_x$, $\left(\frac{\pi}{2}\right)_y$, and $(\pi)_x$ pulses (b, c, and d, respectively).

The most common and simple pulse sequence is to generate primary a echo or Hahn-echo, which is also used as a read-out part in longer sequences (Fig.11) [43, 45].

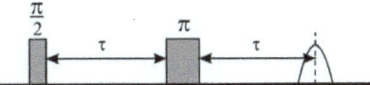

Figure 11. The Hahn-echo. Pulses at distance τ are depicted by grey rectangles, hollow Gaussian represents the Hahn-echo.

The net magnetization vector in equilibrium state (parallel to the direction of the applied magnetic field, $\vec{M} \parallel \vec{z}$) is flipped by the first $\frac{\pi}{2}$-pulse in the *xy*-plane (Fig. 10b). During evolution time τ spins, because of different resonance frequencies, dephase in the *xy*-plane and reach the fenced out state. The subsequent π-pulse inverts the spins where they now start to re-phase. After the second evolution time τ the spins are back in phase along *y*-direction. The signal observed in the *y*-direction is called the Hahn-echo.

2.2.2 Double electron-electron resonance

The system of two coupled electron spins (S_A and S_B) in an external magnetic field can be described by the Hamiltonian which consists of terms arising from electron Zeeman interaction of each spin with the magnetic field as well as exchange and dipolar interaction between those spins, respectively [43]:

$$H_0 = H_{EZ} + H_{ex} + H_{dd} = \omega_A S_z^A + \omega_B S_z^B + J \mathbf{S}_A \mathbf{S}_B + \omega_{dd} S_z^A S_z^B \qquad (3)$$

Here, the exchange and dipolar interaction terms depend on the distance between the interacting spins and thus are of mean for distance measurements by EPR. At short interspin distances (r < 2 nm) the dipolar broadening is detectable in cw spectra [46]. However, at longer interspin distances (r > 2 nm) the contribution from dipolar interaction becomes weaker and should be separated from other interactions in order to be analyzed. This becomes possible in pulsed experiments according to following considerations: The evolution arising from H_{EZ} can be refocused in a simple primary echo experiment (Fig. 11) and consequently terms containing ω_A and ω_B can be dropped off. The exchange interaction decays very fast with increasing interspin distance and is normally negligible at r > 1.4 nm. Finally, the only remaining term of the spin Hamiltonian is that concerned with the dipole-dipole interaction $\omega_{dd} S_z^A S_z^B$.

Double electron-electron resonance (DEER or PELDOR) [47] is a pulse EPR technique that allows for separation of the above mentioned dipolar term from the spin Hamiltonian and, consequently, allows for extraction of distances and distance distributions in the range of 1.5-8 nm. The frequency of the dipolar coupling is related to the interspin distance according to the following equation [48]:

$$\omega_{dd} = \frac{\mu_0 g_A g_B \beta_e^2}{4\pi\hbar} \frac{1}{r_{AB}^3} (3\cos^2\theta - 1), \qquad (4)$$

where μ_0 is magnetic permeability, g_A and g_B are g-values for both spins, β_e – Bohr magneton, r_{AB} and θ – interspin distance and the angle between the interspin distance vector and the direction of the applied magnetic field, respectively, as depicted in Fig. 12a.

The DEER technique exploits two different microwave frequencies to selectively address two fractions of spins. Electron spins in the one part of the EPR spectrum are observed (observer frequency) while pumping at the other part of the spectrum (pump frequency). The first type of this experiment was introduced by Milov *et al.* and consisted of three pulses. There was a primary echo pulse sequence on the observer frequency and a single π-pulse at the pump frequency, where position of the latter was varied (Fig. 12c) [47].

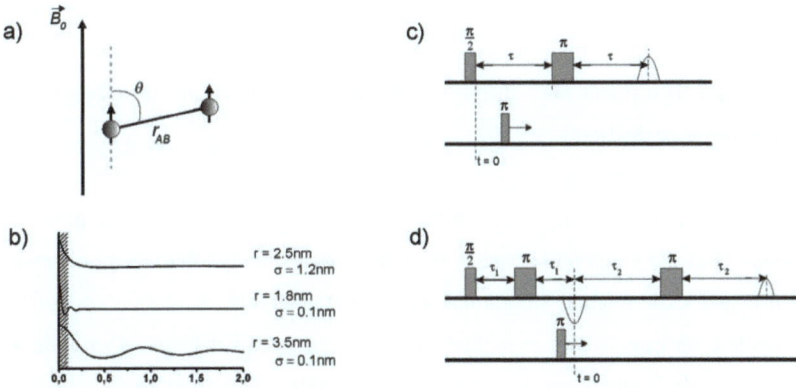

Figure 12. DEER. (a) A pair of dipolar coupled spins in the external magnetic field. (b) Simulated DEER time-traces for Gaussian-like distance distributions with different centers r and widths σ (HWHH). Black-lined area represents a dead-time region. Simulations were performed using a MATLAB-script kindly provided by the research group of Prof. Dr. G. Jeschke. (c) Three-pulse DEER sequence and (d) four-pulse DEER sequence.

In such an experiment the intensity of the echo at the observer frequency is monitored as a function of the time position at which the pump π-pulse is applied. This dependency (Fig. 12b) is called a dipolar evolution curve. The dipolar evolution curve is characterized by the modulation depth λ – the difference in echo intensities at the times where this echo is maximal ($t = 0$) and minimum without any oscillations (t is virtually infinity). For the pair of coupled electron spins the inversion of the pump spin leads to changes in the local magnetic field at the position of the observer spin. The observer spin evolves then in the changed local magnetic field and gains the phase $\omega_{dd}t$. Subsequent movement of the pump spin on the timescale t (Fig. 12c, d) results in a cosine modulation of the echo intensity at the observer frequency, which is given by:

$$V_{3pDEER}(t) = \cos(\omega_{dd}t) \qquad (5)$$

The dipolar evolution curve from real sample, however, has some distinct features different from simple cosine modulation. Among them are decay of the signal due to instantaneous diffusion [49] and damping of the modulations for broad distance distributions [50].

There is a complication in this experiment which is due to existence of the dead-time t_d, a time delay between the first π/2 observer pulse and the pump π-pulse. Thus the pump pulse cannot be applied immediately after the π/2 observer pulse and experimental limitations arise, as the dipolar evolution cannot be recorded starting from $t = 0$. For samples with broad distance distributions, where no oscillations can be seen and for samples with short or intermediate distances (r < 2.5 nm) three-pulse DEER is not favorable because all information on dipolar coupling decays within t_d (Fig. 12b).

This problem can be solved by extending the existing three-pulse DEER with the fourth pulse (Fig. 12d) [50]. The purpose of the additional π-pulse at observer frequency is to refocus the primary echo. Thus the intensity of the refocused echo is an observable in the four-pulse DEER experiment, while the position of the primary echo defines $t = 0$.

Longer distances require to be measured over longer evolution time t which, as being connected to τ_2 (Fig. 12d), is dependent on the phase memory time T_M. Lowering of the temperature can prolong T_M, as well as changing protonated solvent to a deuterated one. The DEER experiment also requires a sample in a motional restricted state (amorphous or crystallite) to prevent averaging of the dipolar coupling [51].

2.3 Nitroxides as spin labels

EPR can be utilized even to study initially diamagnetic biological systems by observation of paramagnetic species that are introduced externally as spin labels or spin probes. The difference between these two groups is in the way they are bound to the system under investigation. Those paramagnetic species which are covalently attached to a molecule under investigation are referred to as spin labels while spin probes can be bound by electrostatic, hydrogen bond, hydrophobic and other forces or remain unbound at all [52].

Generally, each species that carries unpaired electron, belonging either to a transition metal ion or to an organic radical, can be used as a spin label, but it is nitroxides that have emerged as the most popular and frequently used ones. Organic radicals are preferred over transition metal ions due to their narrow EPR spectrum and longer relaxation times. And among organic radicals, being mostly high-reactive, nitroxides have become commonly used spin labels due to their relative chemical inertness and unusual stability [53].

Nitroxides are organic radicals of the general structural formula as depicted in Fig. 13a where R^1 and R^2 may be any substituents and the unpaired electron of the N-O group is delocalized between nitrogen and oxygen atoms to approximately 40 % and 60 %, respectively. Some representative stable nitroxides are shown in Fig. 13.

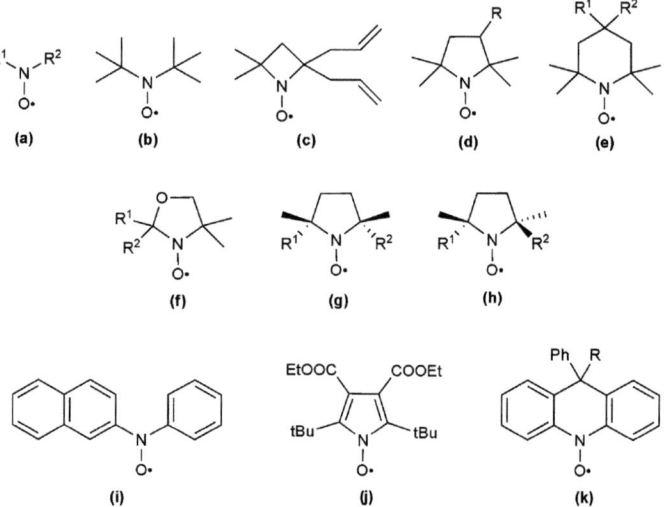

Figure 13. Nitroxides. (a) General formula. (b) DTBN, (c [54], d and e) four-, five- and six-membered ring nitroxides, respectively, (f) doxyl, (g, h) azeothoxyl, (i [55], j [56] and k [57]) aromatic nitroxides with sp²-hybridized α-carbons.

Stability of nitroxides depends primarily on their structure and varies from that of short-living species to persistent radicals which can be stored and handled with no more precautions required for other organic compounds. In particular, structural arrangements around the N-O group have the greatest influence on the stability. As it is evident from Fig. 13, all stable nitroxides are secondary amine N-oxides. Further, they also bear no H-atoms but rather bulky substituents on the α-carbons (carbon atoms which are bound to the nitrogen of the N-O group) which build a steric hindrance towards dimerisation [58]. An absence of H-atoms at α-carbons impedes also a disproportionation reaction which otherwise decreases stability of a nitroxide and leads to formation of nitrone and N-hydroxylamine (Fig. 14) [59].

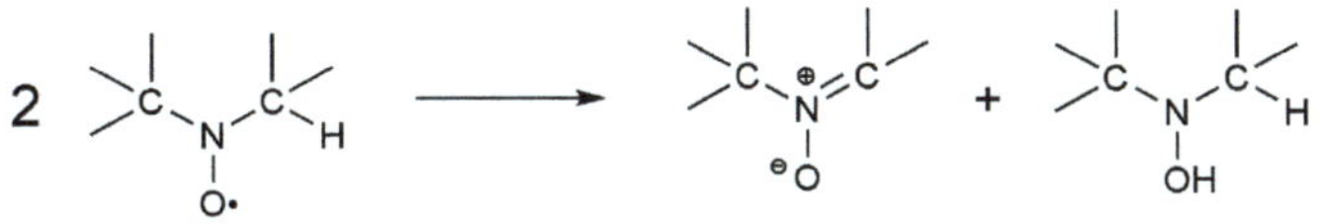

Figure 14. Disproportionation reaction scheme of nitroxides with an H-atom at α-carbon.

Closing the N-O group in a ring does not affect the stability as far as it concerns storing and handling but increases nitroxide's resistance towards chemical reactions (Section 3.2). The nitroxides with α-carbons if they are in a sp^2-hybridized state – that is, being of aromatic origin, – owe their stability also from delocalization of the unpaired electron through conjugation. However, the most commonly utilized spin labels are those on the basis of five- and six-membered ring nitroxides (Fig. 13d-f) [58] as they are molecules of small size and simple structure which allow a variety of possible chemical modifications and have a relatively simple EPR spectrum.

The unpaired electron of the N-O group interacts with the magnetic moment of the nitrogen nucleus ^{14}N (I = 1) thus giving rise to hyperfine splitting of each Zeeman level into three (Fig 15a). Consequently, and according to selection rules, there are three EPR transitions. For nitroxide based on structures with sp^3-hybridized α-carbons (Fig. 13b-h) there is no further delocalization of the electron through conjugation and the EPR spectrum of such a nitroxide spin label consists of only three lines. Its spin Hamiltonian is given by:

$$H_0 = \beta_e \tilde{\mathbf{B}}_0 \mathbf{g} \mathbf{S} + \tilde{\mathbf{S}} \mathbf{A} \mathbf{I}, \qquad (6)$$

where g and A are represented as tensors and reflect the anisotropic nature of both electron Zeeman and hyperfine interactions.

Anisotropy of g-values is rather small (as expected for organic radicals where contributions from the orbital angular momentum are small too) and is fairly unresolved at X-band. As determined from crystal studies of di-*tert*-butylnitroxide DTBN (Fig. 13b) the g-tensor has values [2.0088 2.0067 2.0027] [60]. The principal axes of both g- and A-tensors coincide: the z-axis is parallel to the 2p π-orbital of the nitrogen atom and the x-axis lies along the N-O bond and points in the direction to the oxygen atom (Fig. 15b) [61].

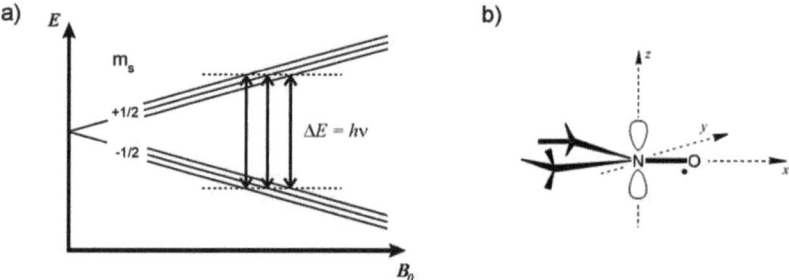

Figure 15. Hyperfine interaction. (a) Splitting of electron Zeeman levels upon interaction with nitrogen nucleus. (b) Principal axes of g- and A-tensors in a nitroxide.

Anisotropy of hyperfine interaction is more pronounced at X-band than the g-anisotropy. The A-tensor possesses almost axial symmetry: $A_{xx} \approx A_{yy} = A_\perp$ and $A_{zz} = A_\parallel$. The hyperfine splitting is much larger in z- than in x- and y-directions (compare $A_{zz} = 0.4-0.5$ mT and $A_\parallel = 3.3-3.7$ mT). The parallel component of the hyperfine tensor is well resolved and is sensitive to the environment of the nitroxide; its value increases in media of higher polarity. The separation between the two external peaks in the three-line EPR spectrum corresponds to a double value of the hyperfine splitting in the z-direction ($2A_{zz}$). Thus a nitroxide spectrum is typically spanned over approximately 200 MHz [62].

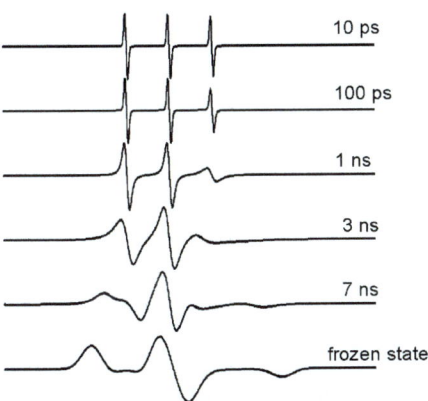

Figure 16. Simulated cw spectra in X-band for a nitroxide having different rotational correlation times. The spectra were simulated using *EasySpin* software package [63]. Following parameters were used for the simulation: g = [2.0088 2.0067 2.0027] and A = [12 12 108] MHz. The rotational correlation time to the each spectrum is given in the figure.

In particular at X-band frequencies, hyperfine anisotropy determines the line shape of the spectrum. Rotation of a nitroxide molecule in solution causes partial or full averaging of the A-anisotropy and additionally influences the spectral lineshape. EPR is thus sensitive to motions with rotational correlation times $\tau_R = 10^{-11}$-10^{-8} s (Fig. 16). In the case of fast rotation the spectrum can be analyzed with Redfield theory and τ_R can be extracted [64]. In this situation hyperfine interaction is completely averaged $A_{xx} = A_{yy} = A_{zz} = a_{iso}$ and the respective parameter can be read out directly from the spectrum as the separation between two adjacent EPR lines. For the case of slow tumbling or frozen spectrum ($\tau_R > 10^{-8}$ s) the full Liouville equation has to be solved [65]. If a nitroxide is rigidly attached to a macromolecule, conclusions about mobility of this macromolecule can also be driven [66].

2.4 EPR on spin labeled DNA
2.4.1 Site-directed spin labeling of oligonucleotides

Nucleic acids, if labeled with nitroxides, can be studied by EPR in terms of their organizational and dynamic properties. Deliberate incorporation of a nitroxide spin label in a nucleotide sequence can be achieved by site-directed spin labeling (SDSL) approach [67] and allows for observation of EPR spectral characteristics of the labeled region or those associated with it. Spin labeling, in general, requires some specific functional groups to which a nitroxide can be coupled. The challenge in spin labeling of nucleic acids is that these macromolecules, being essentially a biopolymer, possess only four different nucleotide building blocks which are connected in a sequence with no specific groups. Nevertheless, nucleic acids can be labeled at each internal site of a nucleotide: at the phosphorodiester group, at the nucleobase or at the sugar moiety. The latter is not used for DNA but only for RNA, because putting any substituents in the 2'-position of the deoxyribose (DNA) converts it automatically to a modified ribose which is a RNA component [68].

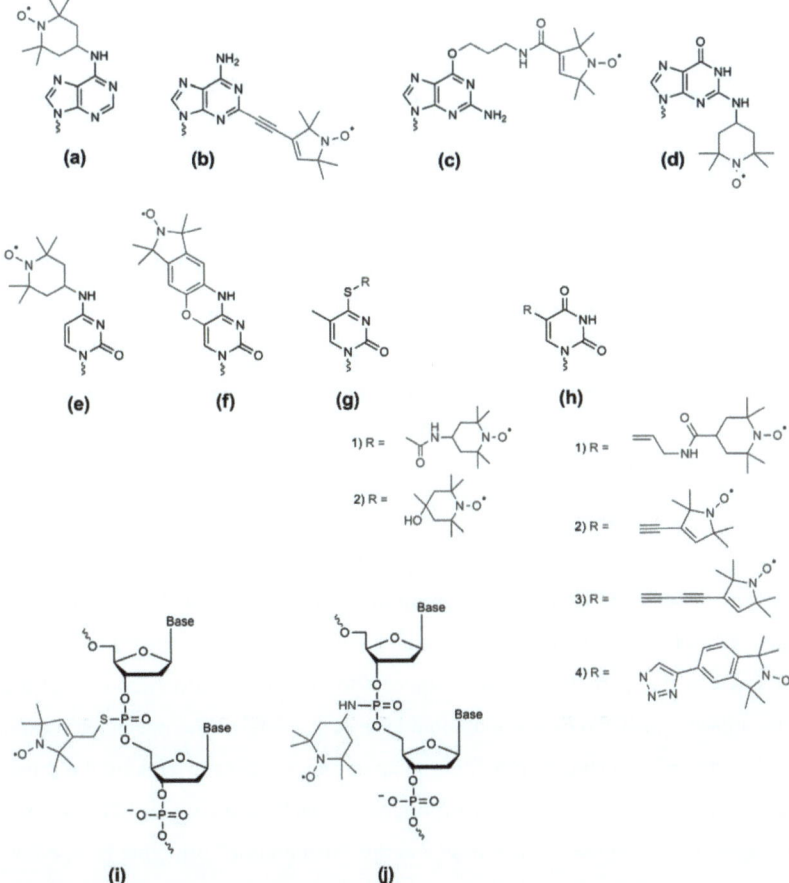

Figure 17. Spin labeling of oligonucleotides. Labeling of nucleobases: (a, b) adenine analogs, (c, d) guanine analogs, (e, f) cytosine analogs and (g, h) thymine analogs. (i, j) Labeling of the phosphodiester group.

Current state of the art in spin labeling techniques allows attaching nitroxides to every of four nucleobases: six member ring and five member ring nitroxides are used. Both purine nucleobases A and G can be labeled at the amino group in

6- and 2-positions, respectively (Fig. 17a, d) [69]. After such an attachment of the TEMPO-like nitroxide only one amino proton remains free for Watson-Crick hydrogen bonding. Despite the fact that the nitroxide is bound directly to the amino N-atom and is situated in close proximity to the base-pairing site, the duplex structure of two spin labeled 20 nt long selfcomplementary sequences appears to be unperturbed (CD spectra) but somewhat destabilized as evident from melting temperature which was 5 °C lower in the spin labeled double helix in comparison to unlabeled one [67]. Further representatives of spin labeled purines are a G analog with a nitroxide in the O6-position which, however, lacks H1 amino proton responsible for Watson-Crick hydrogen bonding (Fig. 17c) [70]; and an A carrying the TPA-label connected to the 2-position via acetylenic tether (Fig. 17b) [71].

There were reported also two spin labeled analogs of the pyrimidine nucleobase C. Similar to compounds in Figs. 17a and 17d, the TEMPO-like nitroxide is attached to the amino group in the 4-position of C thus substituting one H atom (Fig. 17e) and forming a spin labeled nucleobase denoted as ^{T}C. The nitroxide in this particular position appears to be very sensitive reporter of the base adjacent to ^{T}C. While in long double helices a single-base mismatch cannot be determined by measuring melting temperature, because it remains the same, EPR study on a sequence with incorporated ^{T}C can be used instead. The ^{T}C may "recognize" the opposite nucleobase and this "recognition" principle is based on the observation of nitroxides motion in cw EPR at room temperature. Different types of hydrogen bonds which appear upon binding of ^{T}C to other nucleobases restrict the nitroxides' rotation in a different degree so that the corresponding spectra of pairs ^{T}C-G, ^{T}C-C ^{T}C-A and ^{T}C-T can be visually distinguished [72].

The second spin labeled analog of cytosine (Ç) was introduced by the group of Sigurdsson (Fig. 17f). This is an extremely rigid molecular fragment where a nitroxide-bearing isoindole moiety is fused to cytosine by an oxazine linkage. It forms stable Watson-Crick base pair with G and can be incorporated in a DNA duplex without destabilizing the structure significantly (melting temperature is 1 °C lower than for unlabeled duplex) [73]. With the minimal degree of flexibility within the duplex, Ç reports directly on the overall dynamics. Moreover, exhibiting dual spectroscopic activity, it can be converted to a fluorophore upon reaction with mild reducing agents (*e.g.* DTT).

There are two types of thymine spin labeled analogs. In one of them the nitroxide is attached via a flexible linker to a sulfur atom of the 4-thiosubstituted deoxyuridine (Fig. 17g). As a consequence of such modification in the pyrimidine ring, the 3-amino proton is absent and the 4-carbonyl oxygen is exchanged to sulfur thus considerably influencing Watson-Crick hydrogen bonding [74]. The second group of spin labeled analogs of T can be obtained by attaching a nitroxides to a 5-iodo-2′-deoxyuridine via acetylenic tether in Sonogashira coupling reaction (Fig. 17h[2,3]). This spin labeling strategy has found broad application in EPR spectroscopy on nucleic acids. The nitroxide in the 5-position does not disturb DNA duplexes, as was concluded from the similarity of CD spectra and only subtle differences in the thermal denaturation profile for labeled and unlabeled duplexes [75]. Having rather short and rigid linker, the spin label possesses only one degree of motional freedom relative to the base and minimal independent motion. The rotational correlation time measured in EPR implies that the nitroxide's motion is highly correlated with the tumbling of the whole macromolecule but also reports on internal collective motion of the spin labeled duplex and DNA base pairs [75, 76]. It allows experimental estimation of DNA duplex persistence length and access to submicrosecond dynamics [76, 77].

To obtain oligonucleotides with spin labels at desired positions in a sequence phosphoramidites with nitroxide-functionalized nucleobases are used as building blocks in automated DNA synthesis. Sequences up to 80 nt long can be achieved by this synthetic approach [78]. In comparison to unmodified sequences, yields of nitroxide spin labeled oligonucleotides are somewhat lower. This is a consequence of reduction which nitroxides undergo being exposed to reagents of the automated DNA synthesis and which converts nitroxides to EPR-silent hydroxylamines [68]. Thus spin labeling efficiency suffers much for multiple spin labeled sequences if they are obtained by solid-phase synthesis. This problem, however, may be avoided if enzymatic synthesis is used. The sequence containing as many as eleven adjacent spin labeled nucleotides was reported to be synthesized employing DNA polymerase [79].

Obtaining of spin labeled nucleotides in a conventional automated DNA synthesis can be simplified if the nitroxides are attached to already prepared sequence, that is, by postsynthetic functionalization. The spin labeled thymidine analog can be obtained under mild conditions and with quantitative yield by click chemistry where 5-ethynyl-2'-deoxyuridine in the oligonucleotide sequence is coupled to an azide-containing nitroxide [80]. The motional freedom of this spin label within a DNA sequence is similar to that of the frequently used TPA nitroxide coupled to 5-iodo-2'-deoxyuridine via acetylenic linker (Fig. 17h[2,4]).

SDSL of DNA can also be performed in a sequence-independent way by targeting the modified phosphorodiester group in oligonucleotides (Fig. 17i, j) [81, 82]. It gives access to study large DNAs by EPR because the phosphorothioate required for the spin labeling can be substituted at each specific site of an over 100 nt long sequence. The spin label at the phosphorus atom does not appreciably perturb the DNA structure hybridized in a duplex but leads to some loss of the negative charge of the DNA strand [81].

A variety of SDSL approaches offers a number of possibilities to incorporate a nitroxide in a DNA sequence. Attachment of a spin label to a nucleotide can be achieved with different motional freedom of a nitroxide and a degree at which it perturbs the initial DNA structure. Thus, the final choice of the spin labeling technique is defined by samples under investigation and problems to be addressed.

2.4.2 DEER studies of DNA

Once a DNA sequence is spin labeled it can be studied by EPR. In particular, a global structure or structural changes of double spin labeled oligonucleotides can be studied by four-pulse DEER (Section 2.2.2). And nitroxides of only one type are required to label both positions in DNA.

The pump and observer spins are selected as parts of the broad EPR spectrum and are excited with pulses at different microwave frequencies. Pulse lengths are kept short in order to have broader excitation bandwidth and to address as many spins as possible. The choice of the pulse length, however, is limited by the condition that excitation profiles of pump and observer pulses should not overlap. That is why for ~200 MHz broad nitroxide spectrum at the X-band π-pulses of 32 ns (observer) and 12 ns (pump) at approximately 70 MHz separation are used [43, 68]. The pump pulse is usually set to the maximum of the spectrum and excites all orientations, while the observer pulses are applied at the low-field edge (Fig. 18).

The modulation of the refocused echo of observer spins under pumping at the other part of the spectrum is recorded in a dipolar coupling curve (Fig. 12b) and contains frequencies of dipole-dipole interactions between two coupled spins:

$$v_{dd} = \frac{52.16}{r^3}(3\cos^2\theta - 1), \tag{7}$$

where θ and r are defined in Section 2.2.2 and 52.16 [MHz nm^3] is the dipolar splitting constant for nitroxides [48]. As each particular v_{dd} represents a particular interspin distance, analysis of all v_{dd} allows for determination of distance distributions.

In the observed dipolar coupling curve the typical values for λ (if 32 ns observer π-pulses and 12 ns pump π-pulses are used) lie in the range 0.3–0.4 [44, 81-83]. Lower values for λ report on smaller fraction of coupled spins which can be a consequence of a partially reduced nitroxides in the double spin labeled samples.

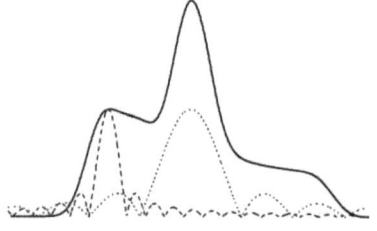

Figure 18. Excitation of the nitroxide spectrum in DEER. Nitroxide absorption spectrum at 45 K (solid line) and excitation profiles of 32 ns pulse (dashed) and 12 ns pulse (dotted) at observer and pump frequencies, respectively.

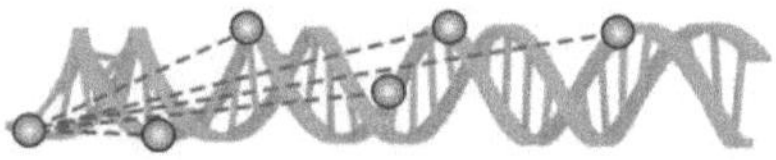

Figure 19. A DNA-based nanometer distance ruler. Summary of different interspin distances within a DNA double helix formed by two single spin labeled oligonucleotides. Circles represent nitroxide spin labels.

The DNA probe in water solution is frozen prior to DEER measurement to avoid averaging of dipolar coupling and to prolong phase memory time T_M. Typically, glycerol is added to the sample: it does not disturb the inherent DNA structure but serves to avoid microcrystallites upon freezing [83].

To evaluate precision at which distances within spin labeled oligonucleotides can be determined in a DEER experiment, a DNA-based nanometer distance ruler was developed. A series of complementary oligonucleotide sequences was designed each of which, bearing initially one nitroxide, yielded upon hybridization a double spin labeled DNA duplex with defined and fixed interspin distance (Fig. 19). Spin labeling was performed by palladium-catalyzed cross-coupling of TPA with 5-iodo-2′-deoxyuridine serving for rigid attachment of the nitroxide to the DNA strand via acetylenic tether. Distances between those nitroxides, as estimated separately for each duplex were 19.2, 22.3, 34.7, 44.8 and 52.5 Å. These results were compared with molecular dynamics (MD) simulations for which the DNA was assumed to adopt a B-form double helix. The DNA was solvated in rectangular box TIP3P water with addition of sodium counterions for compensation of negative charge. MD simulations were carried out over 10 ns for each duplex and distances between O-atoms of the nitroxide groups were determined to be 19.6, 21.4, 33.0, 43.3, and 52.2 Å showing very good correlation with DEER results: $r_{DEER} = 0.99 r_{MD} + 1.1 \text{Å}$ [84].

Another distance ruler was reported to evaluate SDSL distance measurements within a DNA duplex which was spin labeled sequence-independently at the phosphorodiester group. Mean distances calculated from distance distributions of eight DNA duplexes were correlated with those derived from the respective NMR structure: $\langle r_{DEER} \rangle = 1.0 \langle r_{NMR} \rangle - 0.6 \text{Å}$ [81]. Taking into account that derivatization of distance distributions from the NMR structure was based on simple steric exclusion criterion for selecting allowable nitroxide conformations, good agreements with DEER results suggests that the nitroxides have little steric interaction with DNA.

While DEER was shown to deliver precise distances for individual samples, it was important to test its applicability for structurally heterogeneous probes. Here, the same principle of using an oligonucleotide-based distance ruler was applied and a series of DNA double helices was measured having 9, 13, 16, 19 and 23 nucleobases between two spin labels. Corresponding distance distributions for each individual duplex were measured as Gaussians with maxima at 2.8, 4.1, 4.7, 5.5 and 6.8 nm, respectively. Analysis of distance distribution caused by two samples owing 2.8 and 4.1 nm interspin distances is rather straightforward. Two distinct peaks and their ratio can be inspected even visually. Some discrepancies in ratios between measured and expected values can be explained by sample preparation which deals with handling of small volumes of high-viscous solutions and is not very precise. Whereas two distances separated by 1.3 nm can be clearly seen and distinguished in DEER experiment, further mixtures containing equimolar amounts of three, four or five duplexes were tested. Distance distributions from three or four double helices can be fitted with three or four Gaussians, respectively. However, one Gaussian curve of the fit always got shifted from the expected value. The distance distribution from the mixture of five duplexes showed clear shoulders but could not be fitted with Gaussians at expected values [83].

Another potential of DEER is measuring distances between nitroxides with particular orientations to each other. This experiment was reported for the DNA double helix labeled with two rigid Ç-type nitroxides which, due to absence of the tether, keep their orientation in duplex unchanged. The standard set of DEER parameters is used except the position of the observer pulses which, instead of being kept at ~70 MHz offset from the pump frequency, is varied from 90 to 40 MHz in 10 MHz steps. Consequently, distance distributions contain increasing contributions from perpendicular orientations as the frequency offset is shifted from 90 MHz to lower values [44].

Being a method that investigates DNA structure in frozen solution, DEER still can report on DNA dynamics. In particular, due to its ability to measure distributions of distances, it can be utilized to verify conformational flexibility models. The series of double spin labeled 20 nt long DNA double helices was synthesized where separation between two nitroxides was increased in 1 nt step. An extremely rigid spin label Ç was used to exclude any features in distance distributions that may be caused by flexible linker of the nitroxide. The r.m.s.d. values for each distance constraint allow for conclusion about dynamic flexibility of spin labeled region in the oligonucleotide sequence. The r.m.s.d. values for measured distances were plotted against position of the second label in the duplex and compared with r.m.s.d. traces from other conformation flexibility models. The DEER results allowed to exclude those models that claimed helix bending or stretching with constant radius and supported the model of stretching with constant pitch [85].

3. *In-cell* magnetic resonance
3.1 General prerequisites and techniques

For biophysical chemistry the trend to make a step from investigation in buffers that only mimic biological environment towards true "biological milieu" is obvious because it allows studying structures, dynamics and functions under unperturbed and biologically relevant conditions. Spectroscopic methods are the most suitable for observation of biological objects at molecular level due to their inherent non-invasive character. The vibrational spectroscopic techniques, including infrared (IR) [86], Raman [87] and CARS (coherent anti-Stokes Raman scattering) [88] together with fluorescence spectroscopy [89] are well developed for bio-imaging tasks, that is, for observations of cells and tissues at micrometer resolution. In contrast, magnetic resonance techniques (NMR and EPR) are suitable for observations on the molecular level and have all prerequisites to be applied inside cells [90].

Within a cell the object under investigation should possess distinct spectroscopic features that make it "visible" in the complex biological environment. When the object in question does not possess any characteristic spectral features it can be labeled prior to an experiment. While biomacromolecules are isotopically labeled for NMR measurements, EPR requires incorporation of spin labels. For *in-cell* applications conventional nitroxide spin labels can be used as the N-O group was suggested to have low toxicity [91].

There are a number of established protocols that allow deposition of labels or labeled samples inside cells: (i) expression inside cells, (ii) transfection, (iii) uptake and (iv) microinjection. Isotopically labeled proteins for NMR can be expressed inside cells [92], but there are no reports for EPR where biomacromolecules should be marked with spin labels. Biomacromolecules of interest can be transfected into cells if they were encapsulated in a virus [93]. This approach may also work for spin labeled species; however no reports on it appeared up to now. When a molecule can penetrate a cell wall or a cell membrane it can be deposited using cell uptake [94]. For large cells like oocytes from African frog *Xenopus laevis* the microinjection approach can be used [95]. Both uptake and microinjection can be utilized for NMR as well as for EPR studies.

It is noteworthy to mention that while *in-cell* NMR on biomacromolecules is developing since late 1990s [96, 97], the *in-cell* EPR – to the beginning of this work – was only concerned with spin probes for oxymetry [98], study of nitroxide metabolism [99] and application of nitroxides as pH-probes [100].

3.2 Chemical reactions and stability of nitroxides in biological systems

Nitroxides, being stable radicals which persist on air both in powder form and in solution, are exposed to a number of chemical reactions once being put in a biological environment. Already in the first *in vivo* EPR experiment [101] a decrease of nitroxide signal with time was observed. Obviously, the reason for that was a reaction involving the N-O group of the nitroxide which leaded to the loss of a free radical. The nitroxide signal in the EPR spectrum can disappear as a result of conversion of a nitroxide to an EPR-silent oxoammonium salt or hydroxylamine in an oxidation or reduction reaction, respectively (Fig. 20).

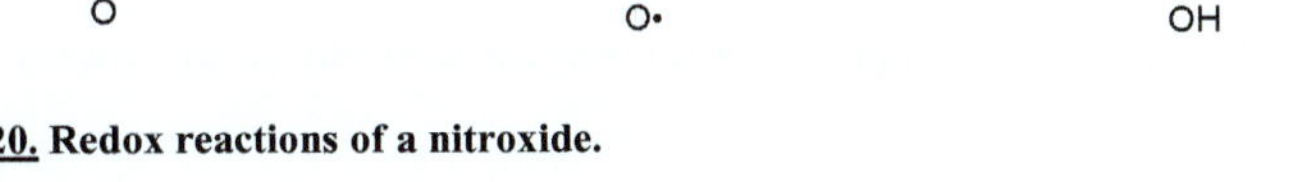

Figure 20. Redox reactions of a nitroxide.

An overoxidation reaction of a nitroxide to an oxoammonium salt can be performed by interacting it with Cl_2, Br_2, moist Ag_2O or CrO_3 which are very strong oxidative agents but are not present in living systems (*e.g.* cells) in fact [58].

In contrary to oxidation, reduction of nitroxides is a much more common reaction which they undergo at rather mild conditions upon interaction with for example ascorbate, dithiotreitol and other thiols. The reduction may have also an enzymatic origin [102]. Having a reduction potential about -150 mV at pH between 6 and 8 (against calomel electrode, $E^0 = -390$ mV) nitroxides may also be reduced by a number of cellular components. Independent on the reducing agent the primary product of the reduction process is a hydroxylamine which can be oxidized back to the corresponding nitroxide by either Fe^{3+} salts [103] or simply by oxygen from air [104].

Reduction by the biologically relevant ascorbate anion is one of those well studied reactions that are also used for testing comparative redox resistance of nitroxides (see below). Kinetics of the nitroxide-ascorbate interaction is normally a second order reaction. In some cases when a reaction goes rather slow and excess of ascorbate should be added it can be treated as one following pseudo-first order kinetics [54]. In living systems the extracellular reduction is due to ascorbate, but intracellular reduction may be caused by enzymes [94].

Thiols, however, do not reduce nitroxides unless other reducing agents are present (*e.g.* superoxide). In this case a superoxide-nitroxide complex is formed which then may either fall apart to initial reagents or react with a thiol yielding hydroxylamine and sulfenyl hydroperoxide RSOOH (Fig. 21) [102]. This reduction pathway was proved to be indeed superoxide-dependent, as in the presence of superoxide dismutase (SOD) the reaction was slowed down to 80%. Additionally, it should be noted that superoxide is produced by several cellular enzyme systems including FAD-containing monooxygenase thus suggesting an enzymatic character to this reduction pathway [105].

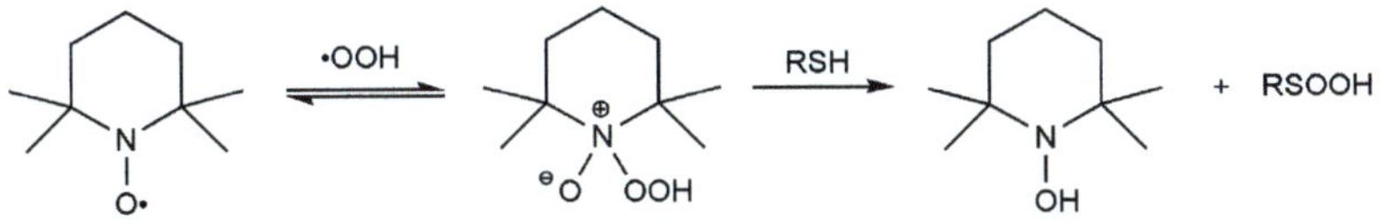

Figure 21. Reduction of nitroxides by thiols.

The cytochrome *p*-450 which function is to transfer electrons is a good candidate for potential reducing agents of nitroxides. However, cytochrome *p*-450 cannot perform reduction alone and needs presence of other reactants [106, 107]. According to the proposed mechanism, at the first stage the free nitroxide radical binds to ferricytochrome *p*-450(Fe^{3+}). This complex accepts electron from cytochrome *p*-450 reductase giving ferrocytochrome *p*-450(Fe^{2+})–nitroxide complex which further undergoes intramolecular electron transfer giving ferricytochrome *p*-450(Fe^{3+}) and corresponding hydroxylamine. To make the whole enzymatic cycle working this system uses NADPH as a source of electrons which converts cytochrome *p*-450 reductase back to the reduced form thus enabling the next turn involving other nitroxide molecule (Fig. 22) [108]. The whole process can be treated with the Michaelis-Menten model. Further, it was shown on the Lineweaver-Burk plot that this reaction can be blocked by known cytochrome *p*-450 inhibitors like carbon monooxide and *n*-octylamine.

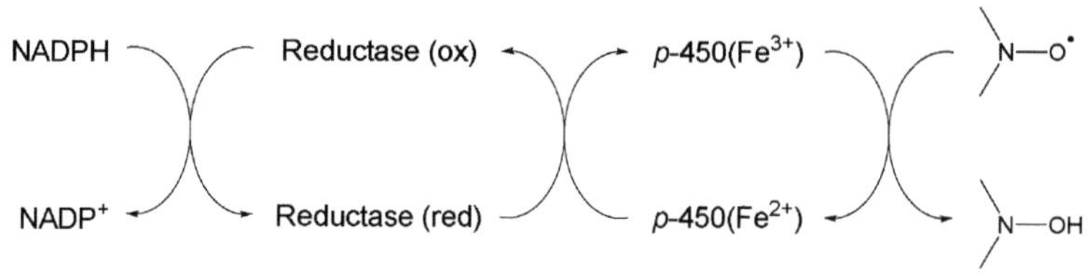

NADPH ⇾ Reductase (ox) ⇽ ⇾ p-450(Fe^{3+}) ⇾ \N—O•

NADP⁺ ⇽ Reductase (red) ⇾ ⇽ p-450(Fe^{2+}) ⇽ \N—OH

Figure 22. Enzymatic reduction pathway scheme of nitroxide by cytochrome p-450.

As the structural features of nitroxides are responsible for their stability (Section 2.3) it is to expect that they may have influence on the chemical resistance of nitroxides towards reduction. The fact if the N-O group is enclosed in the ring or not and the size of the ring are the major factors that define reduction rate in the reaction with ascorbate anion [54]. Thus, in the reaction with ascorbate the five member ring nitroxide (Fig. 13d) has the slowest reduction rate. The same reaction at the same conditions proceeds faster for an open-chain DTBN (Fig. 13b) (10-fold), six member ring (Fig. 13e) (50-fold) and four member ring nitroxides (Fig. 13c) (840-fold). Polar substituents in the ring can additionally accelerate reduction up to tenfold rate [54].

However, the sensitivity of nitroxides towards reduction in real biological systems (*e.g.* cells) is not well understood and depends strongly on both structure of a nitroxide and nature of the cellular environment. From two nitroxides DTBN (Fig. 13b) and doxyl (Fig. 13f) which have the same redox potential only DTBN is reduced by rat hepatic microsomes [109]. For series of five member ring nitroxides, those, having an O-atom in the ring, are selectively not reduced unless mice were pretreated with phenobarbital. At the same time hepatic microsomes from pig do not possess such reduction selectivity and can reduce both DTBN and doxyl nitroxides [102].

The reduction of nitroxides limits however many EPR experiments at room temperature that require longer incubation time to only 30 min or at least 1 hour

[110]. But in spite of the intracellular reduction, nitroxides can still be used and are utilized in oxymetry, as pH-sensitive indicators and probes to observe penetration of the nitroxides inside cells.

II. Experimental

4. Materials
4.1 Chemicals

All chemicals were from Sigma-Aldrich if other not stated. Nitroxides 3-carboxy-2,2,5,5-tetramethylpyrrolidinyl-1-oxy (PCA) and 2,2,6,6-tetramethylpiperidine-N-oxyl-4-amino-4-carboxilic acid (TOAC) were purchased from Toronto Research Chemicals Inc. (Canada). Stock solutions of PCA and TOAC were prepared by dissolving them in Milli-Q water with equimolar amount of KOH.

4.2 Spin labeled DNA

a) Synthesis of spin labeled phosphoramidites

Phosphoramidites bearing six member ring nitroxide 2,2,6,6-tetramethyl-3,4-dehydro-piperidine-*N*-oxyl-4-acetylene (TEMPA) [111] and five member ring nitroxide 2,2,5,5-tetramethyl-pyrroline-1-oxyl-3-acetylene (TPA) [112] were synthesized according to schemes given in Figs. 23 and 24, respectively.

Synthesis of the TEMPA-spin-labeled phosphoramidite was started with commercially available nitroxide which, upon condensation with lithium salt of trimethylsilylacetylene yielded the acetylenic alcohol **2**. The alcohol **2** on methylisation followed by elimination furnished the TMS-enyne **3** and by consequent deprotection yielded the nitroxide **4**. Compound **4** was then coupled with 5-iodo-2'-deoxyuridine via Sonogashira coupling to furnish the nucleoside **5** which on protection with DMT-Cl at the 5'-position and subsequent reaction with cyanoethyl diisopropylphosphoramidochloridite yielded the spin labeled phosphoramidite **7** (Fig. 23).

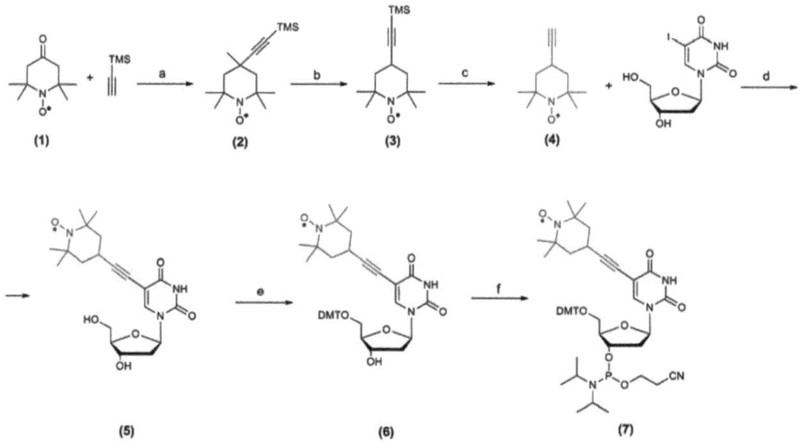

Figure 23. **Synthetic scheme of TEMPA-spin-labeled phosphoramidite.** Reagents and conditions: a) nBuLi, anhyd. THF, 0 °C to rt, 12 h; b) MsCl, Et$_3$N, anhyd. CH$_2$Cl$_2$, 0 °C to rt, 2-3 h; c) TBAF, THF, rt, 1 h; d) Pd(PPh$_3$)$_4$, CuI, Et$_3$N, DMF, rt, 12 h; e) DMT-Cl, DMAP, Py, rt, 12 h; f) (^{i}Pr$_2$N)PCl(OCH$_2$CH$_2$CN), DIPEA, CH$_2$Cl$_2$, 0 °C, 30 min.

Synthesis of the TPA-spin-labeled phosphoramidite was started with 2,2,6,6-tetramethylpiperidine-4-one (**1**) which was converted in four steps to the 3-formyl-2,2,5,5-tetramethyl-1-oxypyrroline (**5**) [113] and subsequently functionalized with acetylenic group yielding compound **7**. Similar to the previous spin labeling scheme, compound **7** was attached to 5-iodo-2′-deoxyuridine via Sonogashira coupling and a resulted nucleoside **8** was converted to TPA-spin-labeled phosphoramidite **10** (Fig. 24).

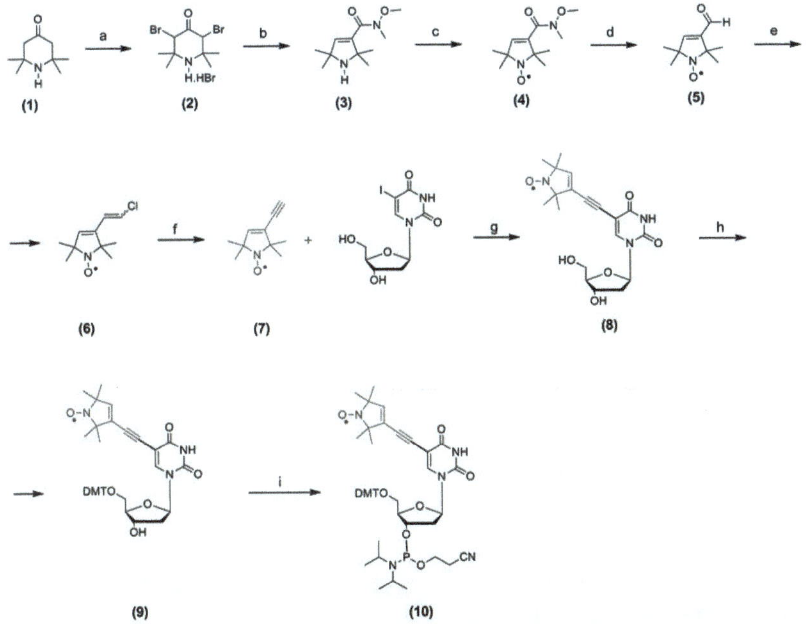

Figure 24. **Synthetic scheme of TPA-spin-labeled phosphoramidite.** Reagents and conditions: a) Br$_2$, AcOH, rt, 6+24 h; b) Me(MeO)NH•HCl, H$_2$O, Et$_3$N, rt, 6 h and 50 °C, 6 h; c) mCPBA, CH$_2$Cl$_2$, 0 °C, 3 h; d) DIBAL, Et$_2$O, -78 °C, 15 min; e) nBuLi, ClCH$_2$PPh$_3$Cl, THF, -78 °C, 1h; f) DIPEA, nBuLi, THF, -78 to 25 °C, 2 h; g) Pd(PPh$_3$)$_4$, CuI, Et$_3$N, DMF, rt, 12 h; h) DMT-Cl, DMAP, Py, rt, 12 h; i) (^{i}Pr$_2$N)PCl(OCH$_2$CH$_2$CN), DIPEA, CH$_2$Cl$_2$, 0 °C, 30 min.

Table 1. Spin labeled oligonucleotides. Spin labeled nucleobases are underlined. DLS – double spin labeled, SSL – single spin labeled sequence.

DNA sequence	Spin label	Short name used in current work
AGGG<u>T</u>TAGGG<u>T</u>TAGGGTTAGGG	TPA and TEMPA	HT-repeat (DSL)
AGGG<u>T</u>TAGGGTTAGGGTTAGGG	TPA and TEMPA	HT-repeat (SSL)
TTGGG<u>T</u>TAGGG<u>T</u>TAGGGTTAGGGA	TEMPA	HT-control (DSL)
TTGGG<u>T</u>TAGGGTTAGGGTTAGGGA	TEMPA	HT-control (SSL)
AGGG<u>T</u>TAGGG<u>T</u>TAGGGTTAGGG TTAGGGTTAGGGTTAGGGTTAGGG TTAGGGTTAGGGTTAGGGTTAGGG	TEMPA	LongHT^{side} (DSL)
AGGG<u>T</u>TAGGGTTAGGGTTAGGG TTAGGGTTAGGGTTAGGGTTAGGG TTAGGGTTAGGGTTAGGGTTAGGG	TEMPA	LongHT^{side} (SSL)
AGGGTTAGGGTTAGGGTTAGGG TTAGGG<u>T</u>TAGGG<u>T</u>TAGGGTTAGGG TTAGGGTTAGGGTTAGGGTTAGGG	TEMPA	LongHT^{mid} (DSL)
AGGGTTAGGGTTAGGGTTAGGG TTAGGG<u>T</u>TAGGGTTAGGGTTAGGG TTAGGGTTAGGGTTAGGGTTAGGG	TEMPA	LongHT^{mid} (SSL)
<u>T</u>ATCGAA and <u>T</u>TCGATA	TEMPA	DNA double helix (DSL)
<u>T</u>ATCGAA and TTCGATA	TEMPA	DNA double helix (SSL)

b) Oligonucleotide synthesis

The syntheses of DNA oligomers were performed on an ABI 394 DNA/RNA synthesizer with commercially available reagents from J. T. Baker and ABI using manufacturer supplied cycles and conditions. The spin labeled DNAs were synthesized on 1.0 μmol scale (1000Å CPG columns) using phosphoramidites with standard protecting groups. The synthesis was performed with standard conditions, except for a longer coupling time for spin labeled phosphoramidite (10 minutes total in several pushes). The DNA oligomers were concomitantly cleaved from the solid support and deprotected by 16 h treatment with concentrated aqueous ammonia at 55 °C. The crude oligomers were concentrated on speed-vacuum and then purified twice by C18-RP-HPLC (0.1M TEAA/acetonitrile, pH 7.0), firstly with "DMT-on" and then with "DMT-off". The purified DNAs were lyophilized and then analyzed by ESI mass-spectromety.

4.3 *Xenopus laevis* oocytes and cell extract

a) Oocytes preparation

Excision and defolliculation of the oocytes was performed as described by Eppig and Steckmann [114]. In summary oocytes were surgically excised from 3-4 year old *Xenopus laevis* females under 0.1% Ethyl 3-aminobenzoate (Agros Organics) anaesthesia. Excised ovaries were placed immediately into 1x MBS medium (10x MBS: 880 mM NaCl, 10 mM KCl, 10 mM $MgSO_4$, 50 mM HEPES (pH 7.8), 25 mM $NaHCO_3$ supplemented to 700 μM $CaCl_2$) [115] and defolliculated by opening the bags with forceps followed by treatment with 0.1% Collagenase Type IA (Sigma C-9891) for 2 hours. Stage VI oocytes [116] were selected and stored at 18°C overnight.

b) Oocyte cell extract

The *Xenopus* cytostatic factor arrested (CSF) oocyte extract was prepared essentially as described by Murray [117]. Briefly: Female frogs were primed by injection with 150 IU of pregnant mare serum gonadotropin. For ovulation female frogs were transferred to MMR medium and injected with 300 IU of human chorionic gonadotropin. Following induced ovulation, oocytes were sorted and washed with MMR medium. Subsequent to de-jellying in de-jellying solution (2% (w/v) cysteine in 1x XB salts (100 mM KCL, 0,1 mM $CaCL_2$, 1 mM $MgCl_2$), pH 7.8) and three washing steps (CSF-XB (1x XB salts, 50 mM sucrose, 10 mM potassium HEPES, 5 mM EDTA, pH 7,7)), activated (dividing) and lysed oocytes were removed and intact oocytes retained. Intact oocytes were then transferred to a centrifugation tube pre-filled with CSF-XB and cytochalasin B. After an initial centrifugation step (2min, 1000 rpm, 4°C) and removal of the supernatant, oocytes were lysed by re-centrifugation (20 min, 10 000 rpm, 4°C). CSF was extracted with a syringe from the middle yellowish layer observed in the centrifugation tube. To control for the integrity of the CSF extract, spindle morphology was observed following addition of $CaCl_2$ and sperm nuclei to a 30-aliquot of the CFS extract [118]. The obtained oocyte extract was frozen and stored at -20 °C until further use (it was taken from the freezer and thawed immediately prior sample preparation).

5. Methods
5.1 CD spectroscopy

5 µM concentration of DNA oligomers was prepared in water with 10 mM Tris-HCl (pH 7.5). Optionally, 100 mM KCl or NaCl were added. Annealing was performed by heating samples to 95 °C followed by slow cooling to 20 °C over 2 hours. In order to reach maximum folding, annealed samples were incubated overnight at 4°C. The CD spectra were recorded on Jasco 715 spectrometer in cuvettes with a 1 cm path length, resolution of 0.5 nm, band width of 1.0 nm and speed of 500 nm/min at 25 °C. Each spectrum was accumulated 5 times and averaged.

For the CD-based kinetic studies, 100 mM KCl solution was added to the pre anealed solution of 5 µM human telomeric DNA oligonucleotides in 10 mM Tris-HCl buffer (pH 7.5). After addtion of K^+, CD spectra were measured every 30 seconds.

For CD-based melting studies DNA samples were prepared as given above and heated from 15 °C to 90 °C with 1 °C/min rate. The CD signal intensity was measured by 290 nm for each 1 °C step. The temperature at the half-intensity of the CD signal was determined as the melting temperature.

5.2 FRET

0.2 µM FRET labeled human telomeric DNA oligomers (FAM at 5′-end and TAMRA at 3′-end of HT DNA) in 10 mM Tris-HCl buffer (pH 7.5) were used to study folding kinetics of G-quadruplexes. 100 mM K^+ solution was added to the pre-annealed FRET labeled DNA oligonucleotides. Immediately thereafter, time-dependent change of fluorescence intensity was measured on Tecan

Infinite 200 PRO microplate reader. Excitation was carried out at 488 nm wavelength and the emission wavelength was 520 nm. Measurements were carried out at every 5 seconds. As control, DNA oligonucleotides annealed in the presence of K^+ were measured.

5.3 Cw EPR

X-band cw EPR masurements were performed on either MiniScope MS200 (Magnettech) or ELEXSYS E580 (Bruker Biospin) spectrometers. Spectra for each sample were acquired at optimal modulation amplitude and microwave power to avoid over-modulation and saturation, respectively.

Reduction kinetics of PCA. Slow reduction kinetics of PCA was followed on the MiniScope MS200 spectrometer operating at X-band (9.4 GHz) equipped with the TC-H02 temperature controller (Magnettech). Each individual sample was prepared by mixing 8 µL of cell extract with 2 µL of PCA and measured in capillaries (Magnettech) varying the nitroxide concentration in the range from 0.04 to 4 mM. For one kinetic curve the sample was monitored over 15 min (1 spectrum/min). Time-dependent automatic acquisition was performed using *AutoIt v3* Software. The amount of nitroxide species in the probe was determined exploiting the height of the low-field peak. All measurements were performed at 18 °C. Each kinetic curve was measured three times.

Reduction kinetics of TOAC. Fast reduction kinetics of TOAC was followed on the ELEXSYS E580 spectrometer (Bruker BioSpin) operating at X-band (9.4 GHz) equipped with an ELEXSYS Super High Sensitivity Probehead, 200 G rapid scan coils and helium gas flow system (ESR900, Oxford Instruments). Each individual sample with TOAC was prepared as follows: A 3 mm OD/1 mm ID quartz tube (Aachener Quarz-Glas Technologie Heinrich, Germany) was filled with 8μL of the cell extract and put in the spectrometer. 2 μL of TOAC were added and mixed using a Hamilton syringe. Reduction kinetics was studied for the concentration range 0.4–20 mM of nitroxide. Kinetic curves were monitored over 2 min in a 2D-experiment (abscissa1 – field; abscissa2 – time) with rapid scan coils (1 spectrum/s). Parameters for rapid-scan experiment were: Sweep ramp-up = 500 ms, Sweep ramp-down = 10 ms, Sweep delay = 10 ms, Time constant = 1.28 ms. The amount of nitroxide species in the probe was determined by double integration of the spectrum. All measurements were performed at 18 °C. Each kinetic curve was measured three times.

Cw spectra at 120 K were acquired either in split-ring MS3 or dielectric MD4 resonator with optimal modulation amplitude and microwave power.

5.4 Pulse EPR: DEER and measurements of relaxation times

a) Sample preparation

For *in-buffer* experiments samples containing spin labeled oligonucleotides in 10 mM Tris-HCl buffer (pH 7.5) and optionally 100 mM of NaCl or KCl were annealed up to 95 °C and slowly cooled down to 20 °C. 40 μL of this solution was mixed with 10 μL of glycerol (20 % v/v) and transferred in the 3 mm OD EPR tube. The final concentration of spin labeled oligonucleotides was 50 μM or 75 μM if not other stated. Samples were shock-frozen in liquid nitrogen prior to measurement.

For *in cellulo* experiments 20 μL of 4 mM stock solution of spin labeled HT-repeat in 10 mM Tis-HCl (pH 7.5) were annealed by heating to 95 °C followed by slow cooling to 20 °C over 2 hours.

For *in-extract* experiments 1 μL of the spin labeled HT-repeat stock solution was added without mixing into a 2 mm ID / 3 mm OD quartz tube containing 24 μL cell extract. After 30 sec the sample was shock-frozen in liquid nitrogen. For consequent measurements the probe was thawed and incubated 11.5 min at room temperature (resulting in a total incubation time of 12 min) and shock-frozen again. The same was repeated for a total incubation time of 20 min.

For *in-cell* experiments oocytes were manually microinjected into the cytosol with 50 nL of the spin labeled HT-repeat stock solution using a Drummond Digital Microdispenser 500 system (Drummond Scientific Company). For *in-cell* DEER on DNA model helix 60 oocytes were used for one measurement. In the case of DNA model helix labeled with TEMPA, each oocyte was frozen in liquid nitrogen in 10 s delay after injection. For life-time studies 25 oocytes were shock-frozen in liquid nitrogen after 0.5, 15, 30, 50, and 90 min, respectively. For distance measurements on HT-repeat 60 injected oocytes were shock-frozen after 15 min. After freezing in liquid nitrogen injected oocytes were lyophilized for 48 h to get rid of ice layer on them which appeared upon shock-freezing. Subsequently, oocytes were transferred into 3 mm ID / 4 mm OD quartz tube. *Note: during all transfer procedures oocytes remain frozen!*

b) EPR measurements

All EPR experiments were performed in X-band using an ELEXSYS E580 spectrometer (Bruker Biospin). For DEER experiments the spectrometer was equipped with a split-ring MS3 (for *in-buffer* and *in-extract* measurements) or a dielectric MD4 (for *in-cell* measurements) resonator and helium gas flow system (CF935, Oxford Instruments).

T_2 relaxation times were measured using the primary echo sequence (Section 2.2, Fig. 11). The pulse lengths were 16 ns and 32 ns for $\pi/2$ and π pulses, respectively. The time separation between pulses t was increased from 200 ns to 1600 ns in 8 ns steps. The relaxation time was determined from the exponential fit of the acquired decay.

The four-pulse, dead-time free DEER sequence is given by: $\left(\frac{\pi}{2}\right)_{obs}\cdots\tau_1\cdots(\pi)_{obs}\cdots t\cdots(\pi)_{pump}\cdots(\tau_1+\tau_2-t)\cdots(\pi)_{obs}\cdots\tau_2\cdots echo$. The pump pulse (24 ns long π-pulse for dielectric resonator and 12 ns long for split-ring resonator) was set to the maximum of the nitroxide spectrum and the observer pulse was shifted 67 MHz higher; π/2 and π pulses at observer frequency were of 16 and 32 ns length, respectively. All samples were measured at τ_2 = 1.5–2 µs and τ_2 = 200 ns. Shot repetition time was set to 4 µs. Typical accumulation times per sample were 10 hours. The DEER time-traces for ten different τ_1-values spaced by 8 ns starting at τ_1 = 200 ns were added in order to suppress proton modulations.

c) EPR data analysis

Processing and distance distribution analysis of all DEER time traces were performed using the *DeerAnalysis2010* software package [119]. Experimental background functions were derived from individually measured DEER traces of corresponding single spin labeled oligonucleotides. Distance distributions from all other DEER measurements were obtained from Tikhonov regularization [120] and subsequently fitted with one or two Gaussian curves. The errors of the distance-distribution parameters were determined by changing those parameters by hand and inspection the agreement of the fit with the experimental DEER time traces. The range of parameters that gives acceptable fits is given as the error margin of the parameters.

III. Results and Discussion

6. Human telomeric DNA in aqueous buffer solutions
6.1 Spin labeling of human telomeric DNA oligonucleotides

The single-stranded overhang at the 3'-end of the human telomeric DNA is a guanine-rich sequence which can form intramolecular G-quadruplexes. A combination of possible strand orientations with a variety of loop types gives rise to manifold of G-quadruplex topologies. Those topologies are energetically very close to each other and even interconversion between different G-quadruplexes may occur. Potentially, any of the known folding types for G-quadruplexes can be adopted by the human telomeric DNA and depends strongly on environmental conditions. The question about the biologically relevant G-quadruplex structure (or structures) still remains open.

While for the human telomeric DNA, as a part of the chromosome, formation of G-quadruplexes in telomerase inhibition experiments was reported, no structural identification of those assemblies has been performed [16]. In contrast, the human telomeric repeat (HT-repeat) d[AGGG(TTAGGG)$_3$] and related sequences containing four GGG tracts and three TTA loops were intensively studied by both NMR and X-ray crystallography in order to determine G-quadruplex folding topologies. Those topologies appeared to be different depending on whether the structure was studied in solution or crystal form and what alkali ions (Na$^+$ or K$^+$) were present in an initial buffer solution. According to high-resolution NMR and CD studies, the HT-repeat in Na$^+$-containing solution adopts only the antiparallel-basket structure. CD spectra of the same HT-repeat in K$^+$-containing solution exhibit features that can be attributed to either (3+1) hybrid or to a mixture of conformations. Noteworthy, only one G-quadruplex type – the parallel-propeller – was determined from single-crystal X-ray studies and no NMR structure in solution is available [20].

Alternative access to study G-quadruplex assemblies in solution can be provided by DEER, a pulse EPR technique, which allows distance measurements in the nanometer range. In a DEER experiment distances between dipolar coupled electron spins are measured [43, 47, 48]. Consequently, for studies of DNA conformations unpaired electrons have to be introduced externally by spin labeling procedure [52, 68]. The distance between those spin labels thus may report on a certain G-quadruplex conformation.

Nitroxides are the most common spin labels and, generally, each of the nucleotides in a nucleic acid sequence can be substituted by its nitroxide-spin-labeled analog (Section 2.4.1). Among three nucleobases that are present in the HT-repeat – adenine, thymine and guanine – all guanines were excluded from spin labeling because folding of the HT-repeat might be disturbed by nitroxides attached to guanines that essentially form a G-core of a quadruplex. Thus, labeling of adenines or thymines in loop regions seemed to be more attractive and safely with respect to preserving an initial G-quadruplex topology. In fact, spin labeling of RNA loops and consequent EPR experiments were performed [121] and no similar studies on DNA secondary structures were reported yet.

Labeling of the HT-repeat was performed using a TEMPA nitroxide attached in the 5-position of the 2'-deoxyuridine (Fig. 25). This spin labeled analog of thymidine with a nitroxide attached to the nucleobase via acetylenic linker was already used for EPR studies on double helical structures

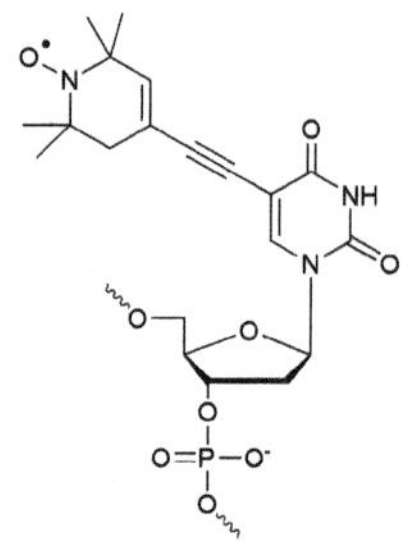

Figure 25. TEMPA spin labeled 2'-deoxyuridine.

of nucleic acids and was reported to be not hampering base pairing and stacking interactions [122].

Table 2. Distances between C5-methyl atoms of thymidines for different G-quadruplex topologies

Position number of thymidines in the d[AGGG(TTAGGG)3]		Distances (nm) in different G-quadruplex topologies			
		parallel-propeller (PDB: 1KF1)	antiparallel-basket (PDB: 143D)	(3+1) hybrid (PDB: 2GKU)	antiparallel-basket with two tetrads (PDB: 2KF8)
5	11	1.7	2.9	2.4	1.0
5	17	2.3	1.1	2.1	1.6
6	11	2.3	2.2	1.5	1.6
6	12	2.8	1.7	0.8	1.8
5	12	2.3	2.4	1.4	1.3
6	17	3.1	0.8	1.9	1.0
6	18	3.8	0.4	1.9	0.5
5	18	3.1	0.8	1.9	1.4
11	17	1.6	2.4	2.4	1.8
11	18	2.2	2.6	2.6	2.0
12	17	2.3	1.7	1.6	1.7
12	18	2.7	2.0	1.7	2.1

Appropriate positions for spin labeled analogs of thymidine within the HT-repeat were found by inspection of available structural data for different G-quadruplexes adopted by oligomers of the human telomeric DNA. Distances between C5-methyl atoms in pairs of thymines were measured (Table 2). The choice of spin labeling positions was guided by two requirements towards the interspin distance. First, the distance between the pair of nitroxides should lies in the range 1.5–3.5 nm. Thus it can be accessed by DEER at even relatively short evolution time of the dipolar coupling curve. Second, distances within the same pair of spin labels have to be characteristic and therefore different for each known G-quadruplex.

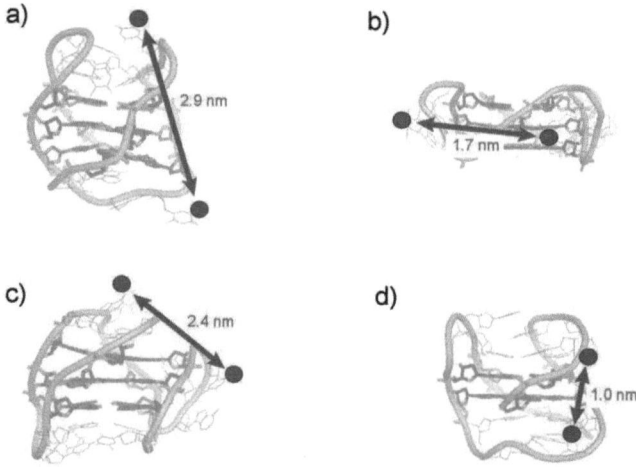

Figure 26. Expected interspin distances for different G-quadruplex conformations. Blue circles represent nitroxide spin labels at positions 5 and 11 in (a) the antiparallel-basket, (b) the parallel-propeller, (c) the (3+1) hybrid, and (d) the antiparallel-basket with two tetrads topologies.

The most suitable positions for spin labeling appeared to be at the first thymidines of the first and the second loops. For the HT-repeat sequence d[AGGG(TTAGGG)$_3$] this corresponds to positions 5 and 11. The same thymidine pairs in related sequences d[TTGGG(TTAGGG)$_3$A] (HT-control) and d[GGG(TTAGGG)$_3$T] are, owing to flanking nucleotides, situated at positions 6-12 and 4-10, respectively. The distance between C5-atoms in those two methyl groups was found to be 1.7 nm in the parallel-propeller structure and 2.9 nm in the antiparallel-basket conformations (Fig. 26a, b; Table 2) which were reported for the HT-repeat. Such a large difference in predicted distances may allow for determination of a certain conformation (or even a mixture of those) in a single experiment. The hybrid structure adopted by the HT-control sequence with the predicted distance of 2.4 nm lie between the parallel and the antiparallel conformations and thus can be distinguished as well (Fig. 26c). The fourth known G-quadruplex – also the antiparallel one – formed by d[GGG(TTAGGG)$_3$T] and possessing only two G-tetrads was reported after spin labeling design of this work has been finished. Fortunately, respective distance between the two C5-methyls appeared to be 1 nm. Such a short distance lies beyond the range accessible in the DEER experiment, but such an interspin distance can be detected by broadening of cw spectra at 120 K (see below).

Site-directed spin labeling of oligonucleotides was performed during a solid-phase synthesis; spin labeled phosphoramidites were used. Presence of incorporated nitroxides in the DNA sequences was detected by cw EPR at room temperature. The spectrum of the HT-repeat single spin labeled with TEMPA at the 5-position features three lines originating from hyperfine coupling to the nitrogen nucleus (Fig. 27a). The lineshape of this spectrum is somewhat different from the spectrum of TOAC nitroxide measured at the same conditions and is a consequence of different rotational motion of nitroxides attached to the oligonucleotide and free in solution (Fig. 27). Quantitative analysis of nitroxides' dynamics in both samples was performed by comparison of their rotational correlation times τ_R. Taking a peak-to-peak linewidth and line amplitudes (line heights) of the cw spectrum in the fast motion limit, the τ_R can be calculated according to the Kivelson equation (8) directly from the experimental data [123]. The rotational correlation time of 988 ps for the single spin labeled HT-repeat is longer than τ_R calculated for the TOAC spectrum (Fig. 27b) by the factor of seven. Consequently, it reflects a larger hydrodynamic radius r_H of the molecule bearing the N–O group in the sample containing the single spin labeled HT-repeat (eq. 9) [124]. The slower motion and the larger hydrodynamic radius of the latter allows for a conclusion that nitroxide in that sample is incorporated in the oligonucleotide sequence and its dynamics is influenced (slowed down) by the overall tumbling of the oligonucleotide in solution.

$$\tau_R = 6.6 \cdot 10^{-10} \cdot W_0 \left[\left(\frac{h_0}{h_{-1}} \right)^{1/2} + \left(\frac{h_0}{h_{+1}} \right)^{1/2} - 2 \right], \qquad (8)$$

where W_0 is the peak-to-peak width of the middle peak, h_i – the height of the peak, i – (-1), 0, (+1) defines low-, middle- and high-field peak, respectively.

$$\tau_R = \frac{\pi \cdot \eta \cdot r_H}{k_B T}, \qquad (9)$$

where η is the viscosity of the medium, r_H – hydrodynamic radius, k_B – Boltzmann's constant, and T – temperature.

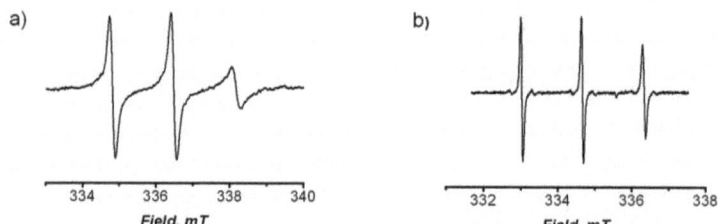

Figure 27. Cw EPR spectra of nitroxides with different rotational dynamics. The spectra of (a) the single spin labeled HT-repeat at position 5 and (b) TOAC correspond to τ_R of 988 ps and 136 ps, respectively. The spectra were acquired at X-band in aqueous solution at room temperature.

While the spin labeled thymidine analog with a nitroxide attached at the 5-postion via an acetylenic linker has already been reported to be non-disturbing in

DNA double helices, its influence on labile and versatile G-quadruplex assemblies is to be investigated. For this goal CD spectroscopy was utilized as an express and independent method which provides some information on G-quadruplex topologies in solution. CD spectra of unlabeled, single and double spin labeled oligonucleotides were measured and compared with each other. No G-quadruplex formation by the HT-repeat is expected in a salt-free buffer. CD spectra of the single and double spin labeled HT-repeat have the same lineshape as the spectrum of the unlabeled sequence (Fig. 28a). The similarity of all the three spectra is clear evidence that incorporated nitroxides do not drive any quadruplex formation of the HT-repeat. The HT-repeat in the presence of Na^+ ions adopts the antiparallel-basket conformation. Both characteristic maximum at 295 nm and minimum at 265 nm were observed in samples with unlabeled and spin labeled sequences (Fig. 28b). It allowed for conclusion that the antiparallel-basket topology remained unchanged upon labeling. The folded HT-repeat in K^+-containing solution was measured by CD as well (Fig. 28c). The spectrum of unlabeled sequence has no features expected for the parallel-propeller structure but rather the maximum at 295 nm and the shoulder at approx. 273 nm. Incorporation of spin labels does not alter any of those features which were observed in the CD profile of the spin labeled oligonucleotides, too. The (3+1) hybrid topology observed for the HT-control in K^+-containing solution exhibits the maximum at 295 nm as the HT-repeat but also much more pronounced shoulder at 273 nm (Supporting materials, Fig. A1). CD spectra show for all those G-quadruplexes that their structures remain unperturbed after attaching of nitroxide spin labels to the oligonucleotide sequences.

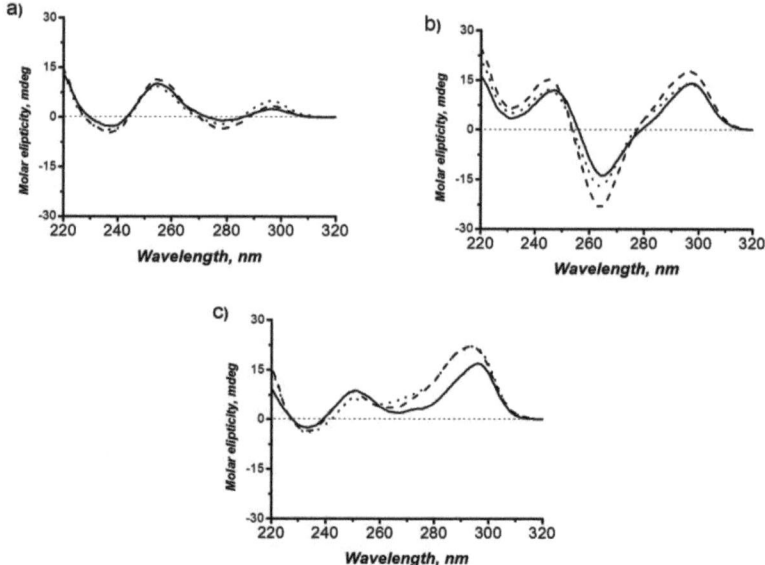

Figure 28. CD spectra of human telomeric DNA oligomers in aqueous solution. Unlabeled (solid), single spin labeled (dotted), and double spin labeled (dashed) HT-repeat in (a) a salt-free buffer, (b) Na^+-containing buffer, and (c) K^+-containing buffer.

The stability of spin labeled G-quadruplexes was investigated by comparing melting temperatures T_m of structures adopted by unlabeled and labeled oligonucleotides. Melting curves were recorded as the intensity of the CD signal at 295 nm upon slow heating of the sample up to 90 °C. The temperature which corresponded to the half of the CD signal intensity was determined as T_m. For the antiparallel-basket quadruplex in Na^+-containing solution the destabilization was ~1 °C per spin label (Fig. 29a). Melting studies of the HT-repeat in K^+-containing buffer were performed on both characteristic wavelengths in the spectrum: at 273 and 295 nm. Melting profiles of labeled sequences recorded at 295 nm – where signal form the antiparallel structure is to be expected – possessed only subtle changes in comparison to unlabeled one. However, destabilization was up to 3 °C. Melting curves recorded at 273 nm were much noisier due to initially low CD signal at this wavelength. There is some destabilization of the structures upon labeling (3 °C per two incorporated spin labels, (Supporting materials, Fig. A2)).

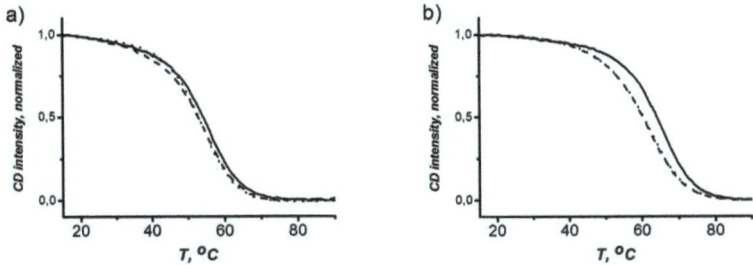

Figure 29. Termal denaturation profiles detected by CD at 295 nm. Unlabeled (solid), single spin labeled (dotted), and double spin labeled (dashed) HT-repeat in (a) Na^+-containing buffer, and (b) K^+-containing buffer.

Thus, incorporation of spin labeled thymidine analogs in loop regions of human telomeric DNA oligomers fulfills the main labeling criteria. The predicted distances between the same spin labeled positions are different for each of known G-quadruplexes and allow for structural identification from a single experiment. Spin labels do not disturb G-quadruplexes as was shown by CD spectroscopy and, finally, folded structures remain stable after spin labeling. The destabilization of several degrees does not exceed those reported for DNA double helices [122]. All mentioned above proves that spin labeled human telomeric DNA oligonucleotides can be used for distance determination and conformational identification of G-quadruplexes by DEER experiments.

6.2 G-quadruplex conformations in Na^+-containing and K^+-containing solutions

Reported up to date DEER-based distance measurements on spin labeled DNA cover only those experiments which deal with double helical structures (Section 2.4.2). Interspin distances and even distance distributions on the nanometer scale are measured with high accuracy and discrepancies between measured and theoretically calculated distances are as small as 0.1 nm [81, 84]. A possibility to predict an interspin distance from the known structure and comparison of that predicted distance with the one measured in a DEER experiment allows for concluding on the particular conformation in solution. The distance distribution originates from the inherent conformational dynamics of a spin labeled macromolecule and from the rotational freedom of a spin label. That is, for rigid double helices the width of the interspin distance distribution is broadened even for those nitroxides attached to the nucleobase via acetylenic linker. Hence, the distance distributions measured for labile structures may be very broad and may complicate structural identification of a spin labeled macromolecule.

First, the antiparallel basket G-quadruplex was studied by DEER. This conformation is known to be adopted by the HT-repeat in Na^+-containing solution. Both CD and NMR studies agree on existence of only this conformer and even full structure for this G-quadruplex is available from high-resolution NMR. For the DEER experiment 50 μM (per oligonucleotide) solution of the double spin labeled HT-repeat in Tris-HCl buffer (pH 7.5) containing 100 mM NaCl was annealed to 95 °C, slowly cooled to room temperature, transferred into a 2 mm (ID) EPR tube and shock-frozen in liquid nitrogen in order to trap the adopted conformation.

The dipolar evolution curve was acquired over 1.4 µs; it contained frequencies of dipolar interaction between coupled electron spins from nitroxides. Those frequencies could be consequently recalculated in respective interspin distances. Pairs of coupled nitroxides which belong to the same oligonucleotide yield **intra**molecular interspin distances. In fact, nitroxides from different DNA sequences may be coupled as well. Those **inter**molecular dipolar interactions also contribute to the dipolar evolution curve of a double spin labeled oligonucleotide. Only intramolecular distances between spin labels are significant for determination of the G-quadruplex topology by DEER. Intermolecular contributions in the dipolar evolution curve thus are attributed to a background and, originating due to spatial distribution of spin labeled molecules in solution, can be either calculated theoretically or determined experimentally.

The single spin labeled HT-repeat with the nitroxide at the 5-position was measured to derive the background function experimentally. Both sample preparation procedure and experimental parameters were kept the same as for the double spin labeled oligonucleotide. The resulted dipolar evolution curve contained only frequencies of intermolecular interactions. Its decay corresponded to the three-dimensional homogeneous background (Supporting materials, Fig. A3) and allowed for excluding of intermolecular G-quadruplexes. This experimentally derived background function was applied to the last two-thirds of the dipolar evolution curve of the double spin labeled HT-repeat in Na^+-containing solution and was used to eliminate intermolecular contributions.

The background-corrected spectrum was analyzed and the distribution of interspin distances was extracted (Fig. 30a, b). Low modulation depth ($\lambda = 0.09$) of the dipolar evolution curve is an evidence for partial reduction of nitroxides in the course of the solid-phase synthesis. Consequently, this reduction decreases labeling efficiency of oligonucleotides and lowers the amount of coupled nitroxides in the sample. However, λ = 0.09 lies above the lower limit of 0.03–0.05 which allows for reliable distance calculation from the dipolar evolution curve [125]. The model-free Tikhonov regularization was applied to analyze interspin distances in this sample and its results fully supported a Gaussian model for this distance distribution (Supporting materials, Fig. A3).

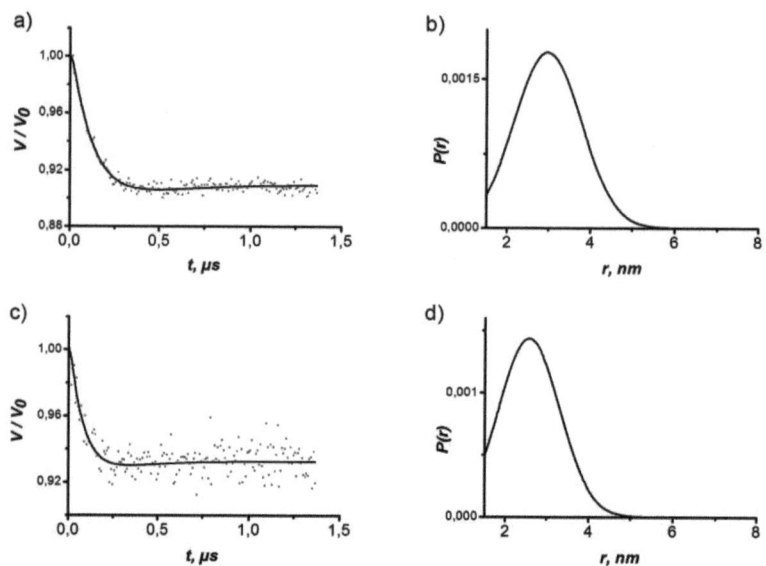

Figure 30. **DEER measurements on single G-quadruplexes.** The HT-repeat in Na^+-containing buffer: (a) dipolar evolution curve (circles) with fit (solid line) and (b) respective Gaussian-like distance distribution. The HT-control in K^+-containing buffer: (c) dipolar evolution curve (circles) with fit (solid line) and (d) respective Gaussian-like distance distribution.

Thus, the distance distribution in the double spin labeled HT-repeat (Na^+-containing solution), if fitted with a Gaussian-like distribution, features its maximum at 3.0 ± 0.1 nm with a width (HWHH) σ = 1.1 ± 0.1 nm. The most probable interspin distance of 3.0 ± 0.1 nm appears to be in very good agreement with 2.9 nm estimated between two C5-methyls from the high-resolution structure of the antiparallel-basket G-quadruplex (Fig. 26a; Table 2). The rather large width of the distance distribution reports directly on the flexibility of loop regions within the G-quadruplex to which nitroxide spin labels are attached.

The main goal for application of DEER to the HT-repeat is to unravel which G-quadruplex structure (or structures) it adopts in the presence of K^+ ions in solution. Rather than operate with theoretically predicted interspin distances it is advantageous to have experimentally determined ones for as many single quadruplexes as possible and to use them as references for evaluation of K^+-dependent folding of the HT-repeat. From the similarity of CD spectra the (3+1) hybrid structure is supposed to be present in K^+-containing solution but a mixture of conformations cannot be excluded as well (Fig. 28; Supporting materials, Fig. A1). A single kinetically stable (3+1) hybrid conformation is formed by the HT-control sequence d[TTGGG(TTAGGG)$_3$A]. Spin labels in the first and the second loop are located at positions 6 and 12 because of two thymidines at the 5'-end of the HT-control instead of one adenosine in the HT-repeat. Prior to calculation of interspin distances the dipolar evolution curve of the sample was corrected for intermolecular contributions using the experimentally derived background function. The distance distribution can be fairly described by a Gaussian at the maximum $r = 2.6 \pm 0.1$ nm with the width $\sigma = 1.0 \pm 0.1$ nm. The value of 2.6 nm lie somewhat higher than the distance between respective C5-methyls was estimated. However, this distance constraint for the (3+1) hybrid topology cannot be confused with the antiparallel-basket conformation and lie far above the expected interspin distance in the parallel-propeller conformation.

Finally, the HT-repeat in K^+-containing solution was studied. The sample preparation was kept the same as for previous samples and included annealing of the double spin labeled oligonucleotide with subsequent slow cooling down to room temperature. This allowed for melting of any DNA structures which might be present in solution into an unfolded state and letting the HT-repeat to fold in the presence of K^+ ions. The adopted conformations in solution were trapped by shock-freezing in liquid nitrogen.

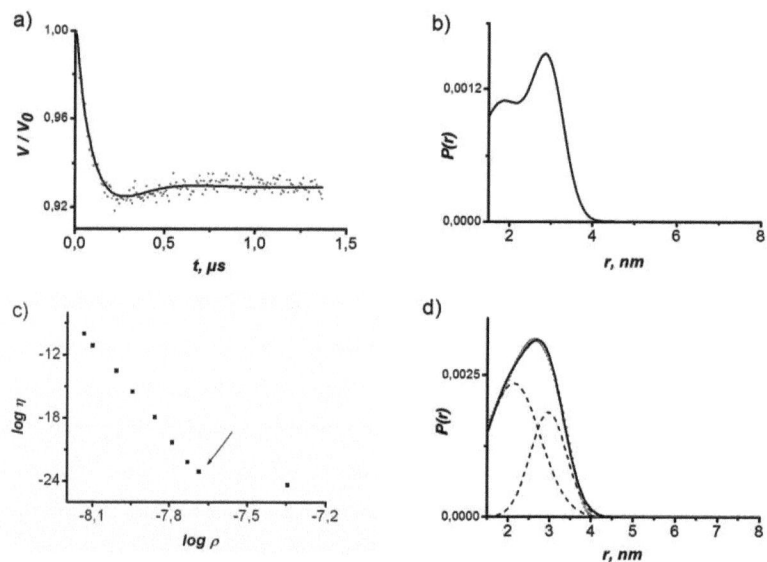

Figure 31. **DEER measurement of the HT-repeat in K^+-containing buffer.** (a) Background-corrected dipolar evolution curve (circles) with fit (solid) and (b) respective distance distribution for the two-Gaussian model. (c) L-curve from Tikhonov regularization, optimal α is marked with an arrow. (d) Respective distance distribution (circles) fitted with two Gaussians (solid) where individual Gauss-curves are depicted as dashed lines.

The dipolar evolution curve was corrected for intermolecular contributions following by a detailed analysis of distance distribution (Fig. 31a). In order to estimate how many conformations are formed by the HT-repeat in K^+-containing solution, model-free Tikhonov regularization procedure was applied.

The optimal regularization parameter α = 10000 was found as the kink of the L-curve (Fig. 31c) displaying a balance between the smoothness of the distance distribution and the accuracy of the fit. The distance distribution envelope could be fitted well with two Gaussians suggesting an existence of two conformations. Thus the two-Gaussian-model was applied and their maxima were extracted to be r_1 = 1.8 ± 0.2 nm and r_2 = 3.0 ± 0.1 nm (Fig. 31b).

Starting points for the analysis of G-quadruplex topologies in K^+-containing solution were interspin distances found for the antiparallel-basket and the (3+1) hybrid structures. The interspin distance for the parallel-propeller structure could not be measured separately, but was expected to be ±0.2 nm around the inter-C5-methyl distance from the PDB structure (Fig. 26b; Table 2). One of the two Gaussian curves centered at of 3.0 ± 0.1 nm describes the distance distribution for the antiparallel-basket conformation which was previously observed in Na^+-containing solution. The width of this Gaussian is narrower than in the respective Na^+-driven structure (Table 3) and suggests a higher rigidity of the antiparallel topology in the presence of K^+ ions. The second peak of the distance distribution is centred at 1.8 ± 0.2 nm and can be fairly assigned to the parallel-propeller structure that was once crystallized from the K^+-containing solution. By determining the area under each Gaussian the ratio of two conformations in solution was determined to be 1:1. Based on the measured interspin distances the (3+1) hybrid conformation was excluded. The antiparallel G-quadruplex with two G-tetrads could be excluded as well by means of no broadening in cw spectra.

Table 3. Interspin distances for G-quadruplexes formed by the spin labeled human telomeric DNA oligonucleotides in buffer solutions.

Sample	Distance constraint r, nm	Width of distance distribution σ, nm
HT-repeat in Na$^+$	3.0 ± 0.1	1.1 ± 0.1
HT-control in K$^+$	2.6 ± 0.1	1.0 ± 0.1
HT-repeat in K$^+$	1.8 ± 0.2 (55 ± 6 %)	0.9 ± 0.2
	3.0 ± 0.1	0.5 ± 0.1

Good agreement between predicted and measured distances within DNA G-quadruplexes proved the applicability of DEER for identification of those structures. Conformational flexibility of spin labeled loops is reflected in the width of measured distance distributions. In spite of the fact that determined distance distributions are rather broad, identification of several G-quadruplexes in one single experiment is possible if difference in characteristic distances exceeds ~0.3 nm. The DEER result of the distance measurement of the double spin labeled HT-repeat in K$^+$-containing buffer showed that there were two co-existing G-quadruplexes one of which was previously crystallized.

6.3 Selective identification of G-quadruplexes within a long DNA sequence

Structural studies of G-quadruplexes of the human telomeric DNA are strongly oriented to define a biologically relevant conformation. Here, one of the issues is a nature of alkali ions present in solution. K$^+$ ions were already shown to play dominant role in the G-quadruplex folding even if 100-fold excess of

Na^+ ions is present [23]. Additionally, higher intracellular concentration of K^+ suggests that exceptional attention has to be paid to those G-quadruplex assemblies that were found in K^+-containing solutions.

However, not only alkali ions determine a G-quadruplex topology. Known intramolecular G-quadruplexes adopted by model sequences of human telomeric DNA show remarkable polymorphism that depends on flanking nucleotides attached to the G-core-forming sequence d[GGG(TTAGGG)$_3$] (Table 4). Those nucleotides define an energetically more preferable structure by formation of hydrogen bonds and π-π stacking of flanking nucleobases on the G-core of a quadruplex. Influence of flanking or, in the broad sense, neighboring quadruplexes on each other have the utmost relevance because the native human telomeric DNA at the telomeric ends of human chromosomes can potentially form a number of adjacent G-quadruplexes.

Table 4. Influence of flanking nucleotides in the human telomeric DNA oligonucleotides on G-quadruplex structures

Sequence	G-quadruplex topology
AGGG(TTAGGG)$_3$	parallel-propeller and antiparallel-basket
TTGGG(TTAGGG)$_3$A	(3+1) hybrid
GGG(TTAGGG)$_3$T	antiparallel-basket with two G-tetrads

In the present study a DNA sequence d[AGGG(TTAGGG)$_{11}$] (denoted as LongHT) which potentially is able to form three intramolecular G-quadruplexes was investigated. The CD spectrum of the LongHT in Na^+-containing solution has a minimum and a maximum at 265 and 295 nm, respectively, and thus provides an indication that also within a triplex of G-quadruplexes only the antiparallel-basket topology is observed for Na^+ (Fig. 32a, black trace). The same oligonucleotide sequence in the presence of K^+ ions features the CD spectrum which is similar to that of the HT-repeat (Fig. 32b, black trace). The maximum at 295 nm and the shoulder at ~273 nm do not allow for deciding on either mixture of the parallel and the antiparallel G-quadruplexes or the single (3+1) hybrid topology.

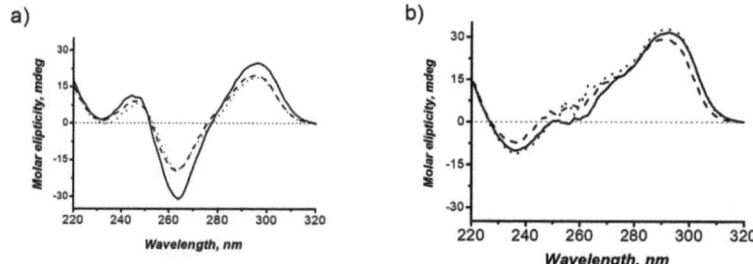

Figure 32. **CD spectra of LongHT oligonucleotides.** Unlabeled (solid), single spin labeled at 5-position (dashed), and double spin labeled at positions 5 and 11 (dotted) LongHT in (a) Na^+-containing and (b) K^+-containing buffers.

TEMPA-spin-labeled nucleotides were incorporated in the LongHT sequence at characteristic positions in the first and the second loop of a G-quadruplex of interest. Double spin labeled LongHT at positions 5 and 11 (Fig. 33a) was used to analyze conformation of the terminal G-quadruplex and the oligonucleotide with nitroxides at positions 29 and 35 (Fig. 33b) was used to investigate the folding topology of the middle part of the sequence which should be surrounded on both sides by two other G-quadruplexes. Thus, in contrary to CD spectroscopy, DEER is to be utilized for observation of the particular structure within a sequence of several quadruplexes and not providing features of the overall folding in the sample.

a) ↓   ↓
AGGGTTAGGGTTAGGGTTAGGGTTAGGGTTAGGGTTAGGGTTAGGGTTAGGGTTAGGGTTAGGGTTAGGG

b)   ↓   ↓
AGGGTTAGGGTTAGGGTTAGGGTTAGGGTTAGGGTTAGGGTTAGGGTTAGGGTTAGGGTTAGGGTTAGGG

Figure 33. Spin labeling of the LongHT. Three G-quadruplex-forming blocks. Arrows show positions for spin labeling in (a) the LongHT^{side} and (b) the LongHT^{mid}.

While nitroxide spin labels in the HT-repeat at positions 5 and 11 point outwards with respect to a G-quadruplex and do not disturb the initial structure (Fig. 26), spin labels in the LongHT sequence may interfere with neighboring G-quadruplexes. Characteristic features of certain quadruplex structures in CD spectra were used for verification wether folding topology of the LongHT remains the same after spin labeling. The LongHT labeled at the side G-quadruplex-forming block (LongHT^{side}) with either one or two nitroxides show identical CD spectra in Na$^+$-containing solution (Fig. 32a). A pair of spin labels in the middle part of the LongHT (LongHT^{mid}) does not alter CD profile as well (Supporting materials, Fig. A6). Similar comparison for K$^+$-dependent folding of the triple G-quadruplexes adopted by unlabeled, single spin labeled and double spin labeled oligonucleotide showed that CD spectra of all samples exhibit the same features (Fig. 32b; Supporting materials, Fig. A6). Consequently, a general conclusion was made that spin labeling of the LongHT DNA neither disturb nor alter G-quadruplex topologies. Identification of those topologies is to be performed by DEER.

Double spin labeled sequences of the LongHT were measured in Na$^+$-containing solution. The sample preparation procedure was kept the same as for the HT-repeat but the concentration of spin labeled oligonucleotides was increased to 75 μM. This increase in concentration together with higher yields of spin labeled oligonucleotides (achieved by double-performed HPLC) served for significantly better signal-to-noise ratio and deeper modulation amplitude.

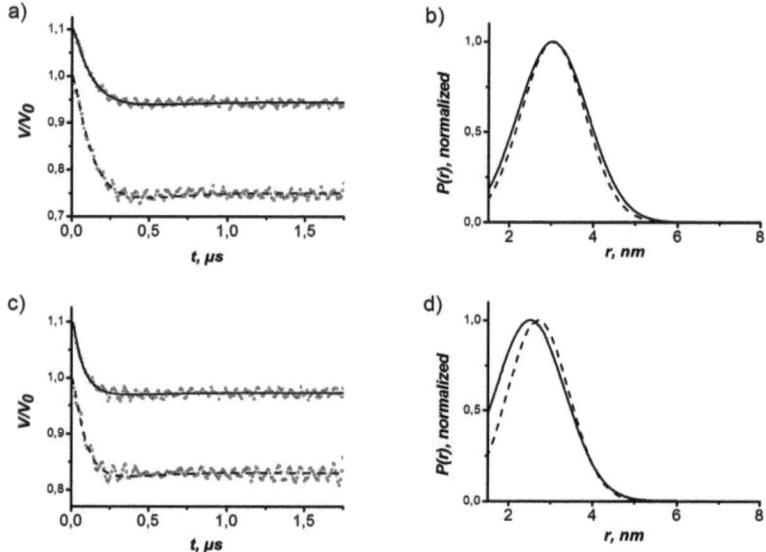

Figure 34. DEER measurements on the Longh HT in (a,b) Na$^+$-containing and (c,d) K$^+$-containing buffers. (a, c) Dipolar evolution curves (circles) with respective fits for LongHT^{side} (dashed line) and LongHT^{mid} (solid line). (b, d) Respective Gaussian-like distance distributions.

Dipolar evolution curves of middle and terminal labeled sequences were first corrected for intermolecular contributions. Here, again experimentally derived background functions from single spin labeled LongHT at positions 29 and 5 were used for LongHT^{mid} and LongHT^{side}, respectively. Resulted dipolar evolution curves, having similar signal-to-noise ratio, are different in modulation depths (λ = 0.15 vs. λ = 0.25 for LongHT^{mid} and LongHT^{side}, respectively). The lower modulation depth for the middle labeled LongHT indicates that less number of spins is coupled with each other. It happens as a result of reduction of nitroxides during the solid-phase synthesis to EPR-silent hydroxylamines. While in the LongHT^{side} TEMPA-spin-labeled nucleotides are added for synthesis of the third G-quadruplex-forming block a pair of spin labels in the middle of the LongHT is longer exposed to reagents of automated DNA synthesis and thus nitroxides are reduced. Consequently, less EPR-active spin labels remain in solution forming a lesser fraction of coupled spins.

Both distributions of interspin distances in the middle and the side double labeled oligonucleotides were calculated assuming a Gaussian model which was shown to be a good approximation and verified with model-free Tikhonov regularization (Supporting materials, Figs. A9-A11). The found maxima of 3.0 ± 0.1 nm for both samples are in excellent agreement with the distance measured for the double spin labeled HT-repeat in the presence of Na^+ ions (Fig. 35a). It allows for conclusion that flanking quadruplexes in the Na^+-containing solution have no additional influence on the overall folding and even within a triplex only one for Na^+ known structure is observed at both side and middle part of the LongHT, that is, the antiparallel-basket (Fig. 35b).

The question about G-quadruplex structures within the LongHT in K^+-containing solution is more intriguing since topological polymorphism of short repeats of the human telomeric DNA is most pronounced in the presence of K^+ cations (Table 4). For topological identification of triple G-quadruplexes in K^+-containing buffer, samples containing 75 µM of the LongHT labeled by pairs of nitroxides at the end or in the middle of the sequence, 100 mM KCl and 10 mM Tris-HCl (pH 7.5) were prepared. The preparation procedure, being similar to previous, included annealing with subsequent cooling down and shock-freezing of the samples. Due to reduction of nitroxides in the course of automated DNA synthesis the same trend like in Na^+-containing samples – a decrease of the modulation depth from $LongHT^{side}$ to $LongHT^{mid}$ – was observed in K^+-containing solutions. Also here the distance distributions between nitroxide spin labels were extracted as those having a form of a Gaussian curve. Interestingly, no distance distributions with two peaks were found. Obviously, there is no mixture of different topologies within one long DNA sequence. The extracted distances were compared to those measured for intramolecular G-quadruplexes adopted by short sequences containing only four GGG tracts (Fig. 35a).

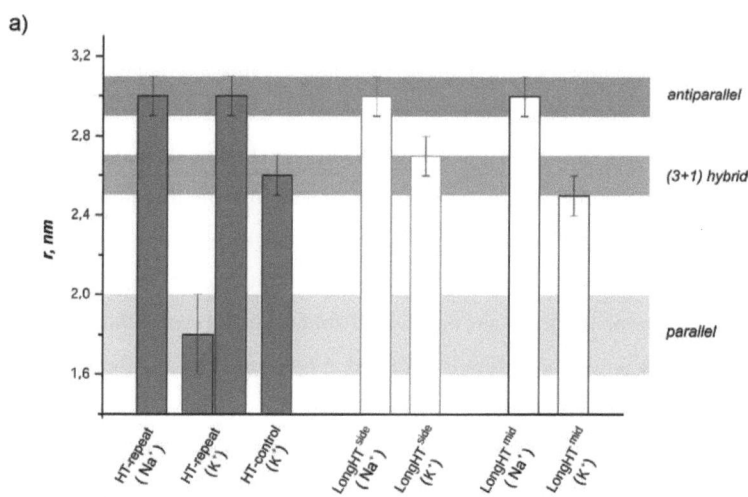

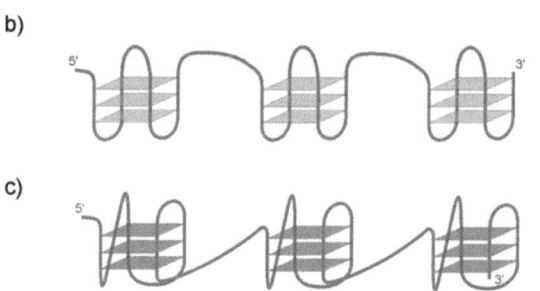

Figure 35. **G-quadruplexes adopted by the LongHT.** (a) Comparison chart of interspin distances measured for the HT-repeat, the HT-control and the LongHT in aqueous buffer solutions. Horizontal bars define regions in which particular G-quadruplexes are expected. Proposed structures for triple quadruplexes formed by the LongHT in (b) Na^+-containing and (c) K^+-containing buffers.

Surprisingly, neither the parallel-propeller conformation ($r = 1.8 \pm 0.2$ nm) nor the antiparallel-basket ($r = 3.0 \pm 0.1$ nm) were found both on the side and in the middle of the LongHT. The extracted distance distributions with most probable interspin distances of 2.5 ± 0.1 nm and 2.7 ± 0.1 nm for LongHT^{mid} and LongHT^{side}, respectively, correspond within the error to 2.6 ± 0.1 nm measured for the (3+1) hybrid structure (Fig. 35a). Some discrepancies in the DEER-measured distances for side and middle quadruplexes can be explained by steric hindrance of spin labels by adjacent G-quadruplexes which force the spin labels to change their spatial orientation and, consequently, leads to slightly different interspin distances.

All dipolar evolution curves recorded for the LongHT-containing samples feature oscillations with frequency of 14 MHz. Those oscillations originate form hydrogen nuclear modulations which were not completely averaged in the DEER measurements. It was not accounted for those oscillations while processing the dipolar evolution curves and extracting interspin distances.

The selectivity of EPR to unpaired electrons in combination with SDSL allowed for targeting and characterization of particular G-quadruplexes within a sequence of those. Each conformation was identified under conditions where steric influence of neighboring G-quadruplexes and eventually forces of π-π stacking or hydrogen-bonding took place. The LongHT in the presence of Na$^+$ ions was shown to adopt the antiparallel-basket conformation which is the same as reported for the HT-repeat. In contrary, neighboring quadruplexes were found to alter folding topology in K$^+$-containing solution. The LongHT sequence was observed to adopt three adjacent (3+1) hybrid structures (Fig. 35c) and neither the antiparallel-basket nor the parallel-propeller was found as it was the case for the HT-repeat.

7. *In-cell* EPR
7.1 Nitroxides inside oocytes and in oocyte extract of *X. laevis*

The non-invasive character of spectroscopy in general is beneficial for studying of living systems; magnetic resonance techniques (NMR and EPR) in particular can be used to characterize whole living systems or selectively their compartments on a molecular level. In fact, while NMR and MRI (magnetic resonance imaging) are utilized for such large living organisms as humans are, *in vivo* EPR is restricted to much smaller living objects (*e.g.* mice) due to limitations in the size of resonator. The main obstacle to extend EPR studies to *in vivo* systems has been the severe dielectric loss and consequent heating that occurs in aqueous samples at conventional EPR frequencies (9.5 GHz, X-band) [126]. However, cells, being essential functional and structural units of any living organism, can easily become objects for EPR studies in the X-band as they fit in a standard EPR tube. Most living systems are diamagnetic or, owing to very low concentration of paramagnetic species, may be assumed to be so. Thus, special attention must be paid to deposition of paramagnetic species inside cells. The choice of a proper way to deliver or to produce radicals into or inside cells depends strongly on type of those radicals and the origin of cells.

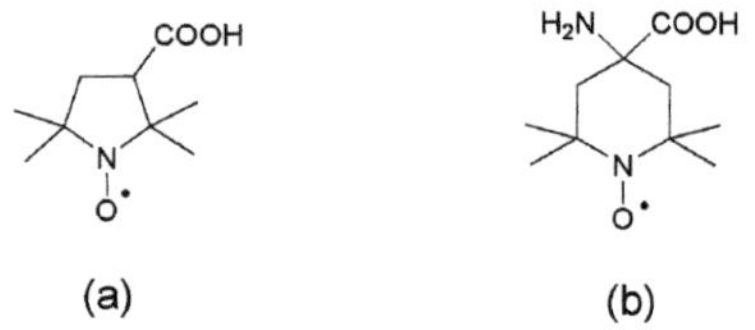

Figure 36. (a) PCA and (b) TOAC nitroxides.

EPR experiments on DNA require spin labeled nucleotide sequences (conventionally, nitroxide spin labels are used). Deposition of a spin labeled oligonucleotide inside cells is concerned with a problem that spin labeled DNA can neither be expressed inside cells nor penetrates through the cell membrane. However, for large cells like oocytes of African frog *Xenopus laevis* (*X. laevis*) the microinjection approach can be applied to transfer spin labeled oligonucleotides in the cytosol. Oocytes have emerged to be useful tools in cell and development biology. They have already been proved to be suitable objects for NMR studies of nucleic acids [28] and potentially could be utilized for *in-cell* EPR. Still, their evaluation for EPR should be performed because special technical and sample preparation requirements must be obeyed.

An oocyte arrested in Stage VI is a spherical cell where nucleus-like conditions preside. Being 1 mm in diameter it contains approximately 1 μL of cytosol. Volumes as high as 100 nL (10 % of the intracellular volume) can be microinjected. Normally, the injection volume is kept between 30-50 nL per oocyte to avoid bursting of the cell.

To evaluate oocytes as possible tools for *in-cell* EPR they were first injected with a solution of a six member ring nitroxide TOAC (Fig. 36b) which is structurally similar to TEMPA spin label used in distance measurements in G-quadruplexes (Section 6). 25 oocytes were injected with 4 mM TOAC in Tris-HCl buffer (pH 7.5). Injection volume of 50 nL per cell resulted in 20-fold dilution of the injected stock solution and TOAC nitroxides were supposed to have intracellular concentration of 200 μM (reduction is neglected). In order to minimize reduction of TOAC inside oocytes, each of 25 oocytes was frozen in liquid nitrogen within 10 s delay after injection. The cw EPR spectrum of the cells in 2 mm ID / 3 mm OD tube was measured at -80 °C and compared with spectra from two reference probes: (i) non-injected oocytes and (ii) buffer solution of TOAC which had similar to the *in-cell* sample, 200 μM nitroxide concentration.

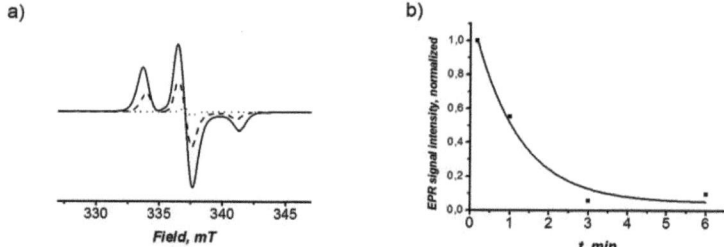

Figure 37. **EPR characterization of TOAC inside oocytes.** (a) Cw spectra at -80 °C of non-injected oocytes (dotted), microinjected oocytes with TOAC (dashed) and TOAC in buffer solution (solid line). (b) Time-dependent decrease of the EPR signal from TOAC inside oocytes with exponential fit.

Both nitroxide spectra from TOAC in buffer solution and inside cells have a lineshape typical for nitroxides in frozen solution (Fig. 37a, black and blue traces). The low-field and high-field peaks from the *in-cell* sample are separated by $2A_{zz} = 200.5$ MHz. It is somewhat smaller than is observed in buffer solution ($2A_{zz} = 213.6$ MHz) and indicates lower polarity of the intracellular environment of oocytes with respect to aqueous solution. The overall intensity of the *in-cell* spectrum, as determined by double integration, contains only 57 % of the *in-buffer* spectrum. While concentration of TOAC inside cells and in buffer was the same this difference might arise from spherical shape of oocytes which prevents them to fill the entire volume of the EPR tube. In spite of the fact that injected oocytes were frozen within 10 s some reduction of nitroxides might occur even within this time window and thus might also contribute to lowering of the overall intensity of the EPR spectrum (see below).

Interestingly, non-injected oocytes feature a signal in EPR spectrum as well. It is a single line at $g \approx 2$ (Fig. 37a, red trace). Its lineshape and position in the spectrum suggests that it can arise from some organic radicals. A control measurement of the cell extract (cell membrane was separated) showed no signal and allowed for conclusion that those organic radicals were present only in the cell membrane. As judged by double integration, oocytes contribute ~2 % to the whole EPR spectrum of microinjected TOAC solution.

In intracellular environment nitroxides are known to undergo reduction which rate strongly depends on both type of cells and type of nitroxides (Section 3.2). The half-life of TOAC in oocytes was determined by measuring intensities of cw EPR spectra from injected oocytes which were frozen after different time delays. In order to avoid any uncertainties connected with the signal produced by oocytes themselves at g ≈ 2 the amplitude of the low-field peak was plotted against incubation time (Fig. 37b). Subsequent fitting of the signal decay with exponential function revealed $t_{1/2}$ = 1.1 ± 0.4 min. Thus, *in-cell* EPR experiments with such nitroxides as TOAC can be performed in frozen state where no reduction occurs, but oocytes have to be frozen fast after injection in order to keep injected nitroxides EPR-active.

In-cell EPR experiments where a sample should be incubated longer at ambient temperature require spin labels which are more resistant towards intracellular reduction. Five member ring nitroxides were recognized to be more stable towards reduction by ascorbate than six member ring analogs [54], but increase in the nitroxides' stability in intracellular environment of oocytes upon changing from six to five member ring has to be verified. The most straightforward experiment is to observe reduction of nitroxides in the cytoplasm-free cell extract. In order to keep intracellular compartments in excess with respect to nitroxides the samples were prepared by mixing the oocyte cell extract with either PCA or TOAC nitroxide (Fig. 36) in 4:1 (v/v) ratio. After fast mixing time-dependent cw EPR spectra were acquired. The amplitude of the low-field peak was used as mean of the amount of EPR-active species and was plotted against time. As it is evident from Fig. 38, 75 % of PCA signal decays within ~5 h at 18 °C. The same amount of TOAC is reduced within two minutes. Aiming to understand and to explain this dramatic difference in reduction rates for both nitroxides in the oocyte cell extract a comparative study of reduction kinetics was performed.

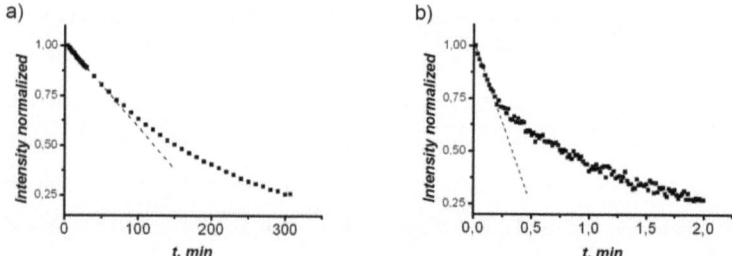

Figure 38. **Reduction of nitroxides in the oocyte extract at 18 °C.** Time-dependent decay of the EPR signal from (a) PCA and (b) TOAC. Dashed lines represent linear approximation of the reduction reaction.

The fact that reduction of PCA runs much slower than for TOAC at chemical similarity of both substances may be explained by highly specific enzymes which could participate in this reduction pathway. The most likely assumption is that PCA does not fit into the cavity of an enzyme as good as TOAC does; that is why the reduction of PCA proceeds slower. Thus, the whole reduction process was assumed to have an enzymatic origin and was treated with the Michaelis-Menten model [127]. According to this model, an enzyme and a substrate (in the present study – a nitroxide) can reversibly react with each other to form a transitional complex which then may either fall apart to the enzyme and the substrate or react further to produce the enzyme and products of the reaction:

$$E + S \underset{k_{-1}}{\overset{k_1}{\rightleftharpoons}} ES \xrightarrow{k_2} E + P \qquad (10)$$

The overall reaction rate is given by:

$$V = \frac{V_{max}[S]}{[S]+K_M} \quad (11)$$

with

$$K_M = \frac{k_{-1}+k_2}{k_1} \quad (12) \qquad V_{max} = k_2[E]_0 \quad (13)$$

where V_{max} denotes the maximal reaction rate, K_M – Michaelis constant, k_i – rate constants and $[S]$, $[E]$ – concentrations of substrate and enzyme, respectively.

Comparative study of reduction kinetics of both PCA and TOAC nitroxides was performed at 18 °C in the cytoplasm-free cell extract of *X. laevis* oocytes which allowed for better manipulation and real-time observations under "near in-cell" conditions. The reduction kinetics was monitored from time-dependently acquired cw spectra at different initial concentrations of nitroxides. The amount of unreduced PCA was determined as the intensity of the low-field peak and plotted versus time. For determination of the non-reduced fraction of TOAC double integration of the whole spectrum was performed. It provided a reliable estimation of the amount of EPR-active species for spectra with broadened lines, as was observed for TOAC concentrations above 10 mM.

Initial reduction rates V were determined as a slope of the linear approximated region of the decay in the EPR signal (Fig. 38, dashed lines). This region does not exceed 20 % conversion. The slower decay of the EPR signal for PCA was monitored with conventional setup for cw EPR (1 spectrum/min). In contrast, the faster decay of the TOAC signal could be monitored with the rapid-scan technique only. Here, less data points were available for initial reduction rates and resulted in larger SEM as compared to PCA (Fig. 39).

A tendency towards saturation of the reaction rates depicted for plots of PCA and TOAC (Fig. 39) indicates that the reaction order changes from one to zero which is characteristic for enzymatic kinetics [128]. Full saturation for PCA was observed in the range 2.5-4 mM of initial concentration. In contrast, no full saturation was achieved for TOAC, albeit the curve does suggest approximation to saturation at concentrations ≥20 mM. Unfortunately, no reasonable experimental characterization appeared possible at concentrations >20 mM as use of highly concentrated TOAC solutions gave rise to extremely broadened EPR lines.

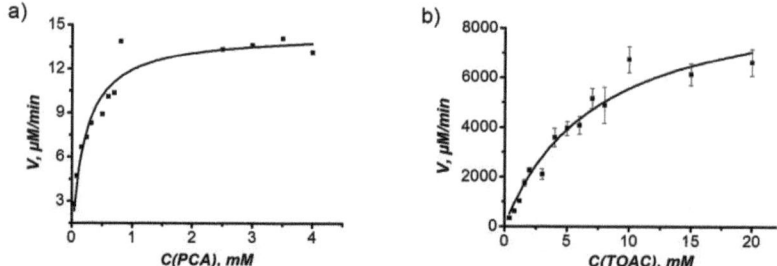

Figure 39. **Dependency of initial reduction rates on the initial concentration of nitroxides.** Michaelis-Menten plots for (a) PCA and (b) TOAC with respective fits.

Kinetic data were fitted according to the Michaelis-Menten model (eq. 11) and respective K_M and V_{max} constants were extracted (Table 5). The V_{max} values for PCA and TOAC were compared with each other. The maximal reduction rate for TOAC appeared to be more that 650 times higher than for PCA. According to eq. 13, V_{max} is proportional to the rate constant k_2 and initial enzyme concentration $[E]_0$. Under the assumption that the same enzymes take part in the reduction of PCA and TOAC, the enzyme concentration can be treated as equal for the PCA and TOAC experiments because the same oocyte extract was used for all measurements. Thus, the ratio between V_{max} for the two nitroxides corresponds to the ratio between their rate constants, that is: $V_{max}(PCA)/V_{max}(TOAC) = k_2(PCA)/k_2(TOAC) = 1/667$.

Table 5. Fundamental constants from the Michaelis-Menten model for reduction of nitroxides in cell extract of *X. laevis* oocytes.

Nitroxide	K_M, mM	V_{max}, µM/min
PCA	0.21±0.03	14.4±0.1
TOAC	7.21±1.50	9600±960

The higher k_2 value for TOAC is also reflected in the Michaelis constant which is determined by the enzyme-substrate affinity and the enzyme-mediated substrate specific reduction rate k_2 (eq. 12). Hence, as no numerical values for k_2 are available also no conclusions can be drawn concerning the affinity between the enzyme and different nitroxides, however, a dominant role of k_2 in the overall reduction rate of nitroxides can be established.

Determination of the maximal reaction rate V_{max} and the Michaelis constant K_M does not allow for identifying a multistep enzyme-mediated reaction. Thus, based on these data a multistep enzyme-mediated nitroxides reduction *in cellulo* cannot be excluded as well. Consequently, the term "enzyme-mediated" includes both reaction types where the reduction is caused either directly by an enzyme or the enzyme should be activated by other reagents in solution.

Stability of nitroxides in the cellular environment of *X. laevis* oocytes depends primarily on the ring size of nitroxides. The reduction process was shown to be an enzymatic one and runs much slower for five member ring PCA nitroxides. Thus, for experiments which demand longer incubation times inside oocytes at room temperature five member ring PCA-like nitroxides should be used. For fast freezing of the sample which allows for stopping of the reduction reaction on its very beginning six member ring nitroxides might be used as well.

7.2 *In-cell* DEER on model systems

A pulse EPR technique like DEER has emerged a useful tool in structural biology. In combination with SDSL it opens access to long-range distances on the nanometer scale thus complementing widely used methods for structural determination – NMR and X-ray crystallography. Aiming to gain structural information of biomacromolecules under physiological conditions, the feasibility of DEER-based distance measurements on a DNA model system inside cells was examined. The experimental design of such a proof-of-principle experiment was concerned with the following issues: (i) choice of a double spin labeled molecule with a well defined interspin distance, (ii) delivery of the spin labeled molecule inside cells, (iii) optimization of the sample preparation procedure and the experimental setup.

DNA oligonucleotides are convenient biological construct to build rigid or semi-rigid 2D and 3D arrangements [3]. In particular, intensive DEER studies on double spin labeled DNA double helices revealed that interspin distances were well defined, could be predicted with high degree of accuracy up to 1 Å and had rather narrow distance distribution if nitroxide spin labels are attached via acetylenic tether. A synthetically obtained double spin labeled DNA double helix cannot penetrate through a cell membrane by itself but may be delivered into cells using a microinjection approach. Large cells like *X. laevis* are suitable for microinjection of volumes as high as 100 nL per cell. High microinjection volumes are required in order to get a sufficient amount of spin labeled molecules which can be detected by EPR. Oocytes are supposed to be frozen after microinjection in order to stop reduction of nitroxides and to trap a DNA conformation.

The model spin labeled DNA double helix was obtained by hybridization of two complementary strands d[TATCGAA] and d[TTCGATA] single spin labeled with TEMPA- labeled thymidine analogs at 5'-end of each sequence (underlined). Positioning of the nitroxide at the end of the strand allows for minimization of the time when the nitroxide is exposed to reducing agents of the automated DNA synthesis; thus higher yield of the spin labeled DNA can be achieved. For the single spin labeled DNA double helix the d[TATCGAA] sequence was hybridized with the complementary but unlabeled strand. Short half-life of six member ring nitroxides in cytoplasmic environment of oocytes should not be a problem in this DEER experiment as no incubation of the DNA inside cells is required and the sample is supposed to be frozen very fast after injection. The interspin distance in this double helix was calculated using SYBYL® software (Tripos Inc.). The methyl groups of respective thymidines were substituted by TEMPA spin labels following by energy minimization of the whole system. The interspin distance in the structure thereafter was estimated to be 3.2 nm.

For microinjection 4 mM stock solution of the DNA double helix in Tris-HCl buffer (pH 7.5) was prepared. 50 nL of the stock solution were manually injected into each oocyte. Taking into account a short half-life of six member ring nitroxides ($t_{1/2} \approx 1$ min) each cell was immediately frozen after the shortest possible time of 10 s. Upon freezing, oocytes have got an ice cover which prevented them to fit into an EPR tube. In order to get rid of that ice cover a careful lyophilization was performed. Cells were taken out of the lyophilizator right away after the ice cover was removed to avoid further water removal from oocytes. Subsequently, microinjected cells were transferred into EPR tubes and stored at –80 °C until measurement.

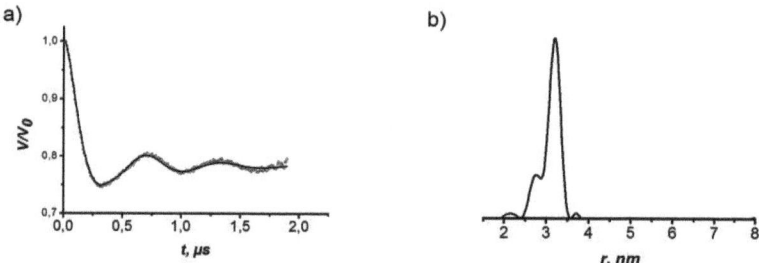

Figure 40. DEER measurement on the model DNA double helix in aqueous solution. (a) Dipolar evolution curve: experimental trace (circles) with the fit for Tikhonov regularization (solid line). (b) The respective distance distribution.

In spite of a normally good agreement between predicted and measured interspin distances within a double helix it is always beneficial to use experimentally estimated distances as a reference for further more complex experiments. Therefore the model DNA double helix was measured in a buffer solution. 1 µL of 4 mM stock solution was mixed with 24 µL of Tris-HCl buffer containing 20 % (v/v) of glycerol and resulted in 160 µM final concentration of the DNA model helix in the sample. The dipolar evolution curve was measured over 2 µs and showed clear oscillations which indicate that the distance distribution is rather narrow (Fig. 40). The Tikhonov regularization (Fig. 40a, blue trace) fitted the dipolar evolution accounting for the oscillations after 0.5 µs. The most probable distance between two nitroxide spin labels was estimated to be 3.2 nm and matched the predicted distance. After obtaining the *in-buffer* reference for the most probable interspin distance and also the envelope of the distance distribution in solution, the *in-cell* DEER measurement was performed.

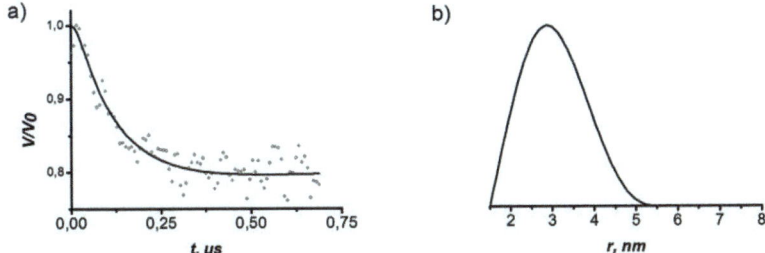

Figure 41. *In-cell* **DEER on the model DNA double helix performed in the split-ring resonator MS-3.** (a) Dipolar evolution curve: experimental trace (circles) with the fit for Tikhonov regularization (solid line). (b) The respective distance distribution.

Microinjected oocytes were measured at the same experimental parameters as the *in-buffer* sample. 30 oocytes were required to fill a 2 mm ID quartz tube for a DEER experiment. Initially low intensity of the refocused echo did not allow for acquiring the dipolar evolution curve over 1.5-2 μs. Instead, the dipolar evolution curve was acquired over 700 ns and corrected for intermolecular interactions. Thus, after the background correction the dipolar evolution curve was analyzed for intramolecular distances. It featured the modulation depth $\lambda = 0.20$ (Fig. 41a) which is somewhat lower than for the *in-buffer* measurement and may be either due to reduction of nitroxides or due to uncertainty from the high noise level. The distance distribution was extracted using Tikhonov regularization. The most probable interspin distance was calculated to be 2.9 nm (Fig. 41b).

The obvious complication in the *in-cell* DEER experiment is a high noise level due to initially low echo intensity. The signal-to-noise ratio may be improved by increase of amount of spin labeled DNA which is detected in course of the measurement. Using stock solutions of even higher concentrations than 4 mM is undesirable. Those concentrations may, from a biological point of view, lead to cell death and, from a point of view of EPR, complicate spectra and whole measurements due to intensive intermolecular interactions and fast relaxation processes. The amount of spin labeled species in the active volume of resonator can be increased using an EPR tube of larger diameter which allows for more dense packing of oocytes in it. While spherical oocytes fill only 57 % of the entire volume of a 2 mm ID tube, ~80 % of the EPR signal is recovered from 3 mm ID tube as a consequence of increasing the filling factor of the resonator.

60 oocytes are needed for one DEER sample to be measured in a 3 mm ID / 4 mm OD tube. Also a dielectric resonator MD-4 with a larger diameter was used instead of a split-ring MS-3. While sample preparation procedure was kept the same the length of the pump π-pulse had to be increased from 12 ns to 24 ns in order to reach a flip angle of 180°. The sample was measured overnight at 45 K. Better filling factor of the MD-4 resonator provided more intensive spin echo and allowed for detection of the dipolar evolution curve over 1 µs. However, the modulation depth of the dipolar evolution curve ($\lambda = 0.11$) is approximately twice as low as in the MS-3 resonator at the same noise level (Fig. 42a). It is a consequence of the two times longer π-pulse used which, having narrower excitation bandwidth, is more selective and excites a smaller fraction of spins.

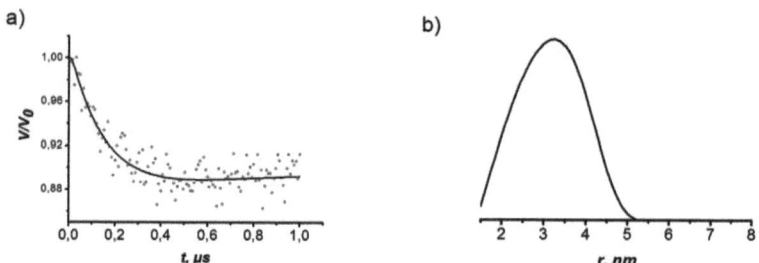

Figure 42. **In-cell DEER on the model DNA double helix performed in the dielectric resonator MD-4.** (a) Dipolar evolution curve: experimental trace (circles) with the fit for Tikhonov regularization (solid line). (b) The respective distance distribution.

The dipolar evolution curve was fitted with Tikhonov regularization resulting in a broad distance distribution with maximum at 3.2 nm (Fig. 42). The most probable interspin distance within the model DNA double helix inside cells matches experimentally obtained distance in buffer very well. The increased width of the distance distribution measured inside cells may originate from low signal-to-noise ratio and thus from uncertainty in derived distances. Also, broader distance distribution may be a result of partial melting of the short double helix (consisting of 7 bp) in cellular environment of oocytes. The results from MD-4 resonator are more promising for *in-cell* DEER because longer dipolar evolution curves could be acquired. Finally, exchange to five member ring nitroxide might also improve signal-to-noise ratio and thus might allow for more precise distance determination.

In parallel to this study a report of Krstić *et al.* appeared describing *in-cell* DEER on nucleic acids as well [129]. Using of five member ring nitroxide TPA as spin label in a 12 bp DNA double helix allowed them for observation of narrow distribution of interspin distances as is characteristic for rigid DNA duplexes. The more stable towards reduction TPA spin label made it possible to measure DEER even after 70 min incubation time. The fact that narrower distance distribution was observed for the more stable 12 bp double helix than for 7 bp one supports our hypothesis that shorter double helix melts in cytoplasmic environment of oocytes and this melting caused broader distribution of interspin distances (Fig. 42b).

Analysis of the above described results for nucleic acids together with pioneering work of Igarashi *et al.* about *in-cell* DEER on proteins [110] allows for deriving general recommendations for *in-cell* distance measurements. Reductive properties of the cellular environment primarily determine optimal conditions for the sample preparation. Regardless the type of biomacromolecules, a protein or a nucleic acid, more stable five member ring nitroxides should be used as they allow for longer incubation of the sample *in cellulo* and easier handling. Attachment of spin labels to biomacromolecules should avoid using S-S bridges which might be reductively cleaved inside cells. In nucleic acid conventional acetylenic linker can be used. However, for proteins the widely used MTSL spin label should be exchanged to for example 3-maleimido-PROXYL which bounds to the cysteine via C-S bond formation. EPR tubes with larger inner diameters (and respective resonators) 3 or 4 mm are recommended to increase filling factor of a resonator and to lower the noise level.

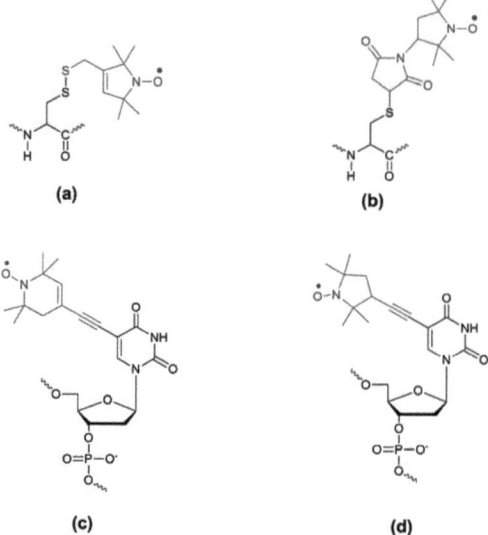

Figure 43. **Strategies for spin labeling of proteins and nucleic acids for *in-cell* EPR.**
MTSL-labeled cysteine (a), 3-maleimido-PROXYL-labeled cysteine (b), TEMPA-labeled 2'-deoxyuridine (c), and TPA-labeled 2'-deoxyuridine (d).

7.3 Conformations of the human telomeric repeat *in cellulo*

Conformations of G-rich DNA sequences are determined by their environment. While intramolecular G-quadruplexes formed in the presence of Na^+ or K^+ ions by DNA sequences with different flanking nucleotides were identified, the G-quadruplex structure of the human telomeric DNA at native conditions cannot be predicted *a priori*. Not a certain environmental factor but rather a combination of many of them defines the folding topology of a G-rich sequence inside cells. Thus, in order to determine conformations of labile G-quadruplexes it should be accounted for effects of molecular crowding and viscosity, presence of small molecules and any biomacromolecules, influence of pH and temperature. NMR, X-ray, CD and recently EPR derived a body of results on G-quadruplex topologies adopted by fragments of the human telomeric DNA (Section 6). Additionally, establishment of *in-cell* DEER for distance measurements on nucleic acids (Section 7.2) suggests that structural investigation of G-quadruplex assemblies may now be performed under intracellular conditions thus accounting for the complex set of environmental factors inside cells.

Regarding the remarkable structural polymorphism of intramolecular G-quadruplexes adopted by HT DNA oligonucleotides, the HT-repeat d[AGGG(TTAGGG)$_3$] appears to be very interesting and intriguing object for determination of its folding topology inside cells. Briefly, the HT-repeat was shown to adopt antiparallel-basket structure in Na^+-containing solution. In the presence of K^+ ions it was found to form 1:1 mixture of antiparallel-basket and parallel-propeller topologies. Initially, only the parallel-propeller was identified from a crystallized sample. Moreover, influence of molecular crowding on the determination of G-quadruplexes was studied. The parallel-propeller structure of the HT-repeat is favored in K^+-containing solution crowded by PEG. It was also observed that other G-quadruplexes formed in the presence of K^+ ions (*i.e.* the (3+1) hybrid and the antiparallel-basket with two tetrads) undergo conformational transition to the more compact parallel-propeller structure which have been firstly determined from X-ray crystallography [25].

G-quadruplex topologies of the HT-repeat in the intracellular environment can be identified in the *in-cell* DEER experiment by measuring characteristic distance constraints between nitroxide spin labels. The experimental design includes the following steps: (i) to deposit the unfolded HT-repeat inside cells, (ii) to incubate the sample to enable its folding, (iii) to shock-freeze the sample in order to trap a conformation or conformations, (iv) to measure characteristic distances between spin labels and (v) to conclude on G-quadruplex structures formed under intracellular conditions.

The HT-repeat was labeled at positions 5 and 11 which provided characteristic interspin distances for identification of G-quadruplexes. Instead of TEMPA which was used for *in-buffer* studies (Section 6) the HT-repeat was labeled with the TPA nitroxide. In the latter spin label the N-O group is enclosed in a five member ring which serves for higher resistance towards reduction *in cellulo* (Section 7.1). Possible deviations in distances measured between TMP spin labels in comparison to TEMPA should be completely compensated by inherent flexibility of the loops which owe to higher degree of freedom than nitroxides attached via an acetylenic linker. The measured distances are supposed to be in agreement with previously estimated within the error of 0.1 nm provided by the DEER experiment itself (Section 6). In fact, from the good agreement between inter-C5-methyl distances and measured interspin distances (Section 6.2) any significant deviation resulting from exchange to TPA nitroxide can be excluded.

According to the experimental design, the spin labeled HT-repeat should be incubated inside cells to form G-quadruplexes. The proper incubation time has to be long enough to provide folding of the HT-repeat and to allow it for reaching an equilibrium state. On the other hand, incubation time should be kept as short as possible to minimize a number of reduced nitroxides and, consequently, to minimize signal losses in DEER.

FRET is a convenient and fast method for observation of G-quadruplex folding [32]. The solution of KCl was added to the HT-repeat labeled with the FRET pair (FAM – donor and TAMRA – acceptor at 3'- and 5'-ends of the sequence, respectively) following by time-dependent measuring of the fluorescence signal at 520 nm. Upon folding of the HT-repeat FAM and TAMRA become closer and energy transfer takes place. The fluorescence intensity at 520 nm decreases with time during G-quadruplex formation and reaches the signal intensity which corresponds to completely folded conformation already within 2 min (Fig. 44a).

CD spectroscopy provides express information on some types of G-quadruplexes in solution. Although only the parallel and the antiparallel topologies may be unambiguously identified CD spectra allow for clear distinguishing between folded and unfolded states. Thus, time-dependent CD spectra were acquired immediately after adding KCl to the solution of the unfolded HT-repeat. In agreement with FRET results, the HT-repeat is folded within approx. 2 min. However, subtle changes at ~273 nm are observed 3 min more. It can be concluded that the HT-repeat reaches an equilibrium state after 5 min incubation time because no further spectral changes were observed (Fig. 44b).

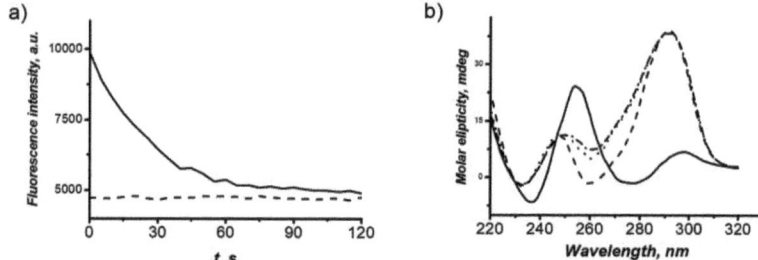

Figure 44. Folding kinetics of the HT-repeat in K^+-containing buffer solution. (a) Time-dependent FRET spectra of the folded HT-repeat (dashed line) and during the folding process (solic line). (b) Time-dependent CD spectra: after 0 min – without K^+ (solid line), 0 min 15 sec (dashed line), 6 min 15 sec (dotted line), 9 min 15 sec (dash-dotted line).

Also the life-time of nitroxide spin labels inside oocytes was tested. Although intensive study of reduction kinetics of the five member ring nitroxides PCA was already described in Section 7.1, reduction of the TPA spin labels has to be investigated in order to choose a proper incubation time for this particular nitroxide. The solution of the HT-repeat, single spin labeled at position 5, was injected into oocytes in order to account for influence of viscosity inside cells and different diffusional motion, rotational mobility and accessibility of a nitroxide spin label bound to a biomacromolecule for reducing agents in a bulk solution. Subsequently, the microinjected oocytes were frozen after different time delays and cw spectra at 120 K were acquired. A typical nitroxide spectrum in frozen state was observed after 10 s incubation (Fig. 45a, green trace). However, the observed nitroxide spectrum is overlapped with a single line at g ≈ 2 which origins from the oocytes. The initially invisible signal from oocytes in the cw spectrum acquired after 10 s delay becomes more pronounced at longer incubation times (Fig 45a, red trace) where intensity of nitroxide spectrum decreases due to reduction reaction *in cellulo*. Thus, the amplitude of the low-field peak which is attributed solely to spin labels was taken as a mean of the amount of nitroxides in the sample and was plotted versus incubation time (Fig.

45b). The decay of the EPR signal was fitted with an exponential function resulting in a time-constant of 29 ± 7 min.

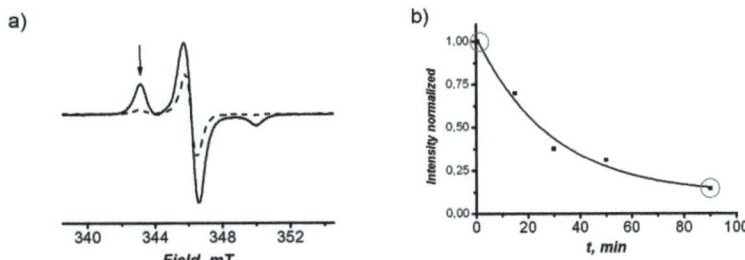

Figure 45. Reduction of the TPA nitroxides within the single spin labeled HT-repeat inside *X. laevis* oocytes. (a) Representative cw-EPR spectra: 30 sec and 90 min after injection (solid and dashed line, respectively). (b) Time-dependent decay of the signal intensity (amplitude of the low-field peak) in the nitroxide spectrum. Points in circles correspond to cw spectra shown in (a).

Folding studies of the HT-repeat performed by FRET and CD in aqueous buffer solutions showed that G-quadruplexes were formed and reached an equilibrium state within 5 min. However, this value provides only a rough indication for the high-viscose cellular environment where all diffusion and thus folding processes may occur slower. Hence, the incubation time of the HT-repeat inside oocytes should be longer than 5 min and is limited by the half-life of the TPA spin labels. The optimal time of 15 min was chosen as the longest one after which many of nitroxides still remain unreduced. It would allow for keeping as much spins as possible at a relatively long incubation time.

For *in-cell* DEER the 4 mM stock solution of the HT-repeat was injected into 60 oocytes (50 nL per cell) resulting in a final effective intracellular concentration of 200 μM. While successful DEER measurements in buffer solutions can be performed already at 50-75 μM concentrations of double spin labeled species [48], the value of 200 μM allows for observation of an intense signal even if the nitroxides get partially reduced. The microinjected oocytes were incubated 15 min at room temperature and shock-frozen in liquid nitrogen. Careful lyophilization was carried out in order to get rid of the ice cover on the microinjected cells without dehydration of the cytosol (see Section 7.2). The sample in a 3 mm ID tube was measured at 45 K overnight. The 1.5 µs-long dipolar evolution curve was acquired and corrected for intramolecular dipolar contributions with a background function derived from the experimentally measured single spin labeled HT-repeat inside oocytes. The usage of experimentally derived background functions is the most reliable for *in-cell* samples where injected spin labeled species do not reach homogeneous distribution over the cell volume within a given incubation time.

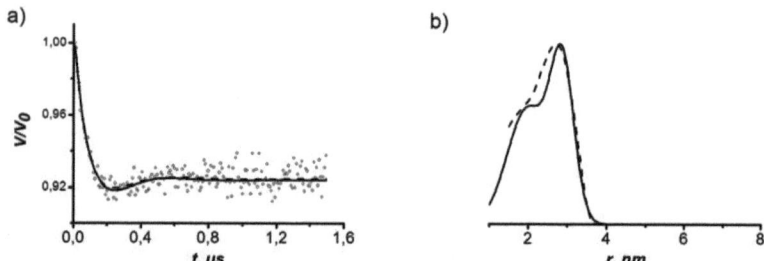

Figure 46. *In-cell* **DEER on the HT-repeat.** (a) Dipolar evolution curve: experimental trace (circles), two-Gauss fit (solid line) and Tikhonov regularization (dashed line) with (b) respective distance distributions.

The resulted dipolar evolution curve (Fig. 46a) featured lower modulation depth as compared to *in-buffer* and *in-extract* experiments (see below). The lower value of λ measured for the *in-cell* sample is mainly due to the longer pump π-pulse (24 ns) used in the setup with the MD-4 resonator. Additionally, lowering of the modulation depth was caused by partial reduction of nitroxides during the incubation of the HT-repeat inside cells at room temperature. The distance distribution obtained by model-free Tikhonov regularization at optimal regularization parameter α = 1000 has two distinct peaks (Fig. 46b, blue trace). Consequent fit of the dipolar evolution curve with two-Gaussian-model revealed distance distribution with two maxima at 2.0 and 2.9 nm (Fig. 46b, red trace; Table 6). These distances, if compared with the DEER results in K^+-containing solution (Section 6.2), can be attributed to the parallel-propeller and the antiparallel-basket G-quadruplexes. The ratio between those conformations is 1:1 and thus the same as in K^+-containing buffer. The distance distribution envelope, however, do not fall complete to zero at distances shorter than 1.5 nm. Unfortunately, an approach to analyze those short-range distances by broadening in cw spectra cannot be applied for *in-cell* samples where two signals – from nitroxides and from cells – are overlaid. The signal from oocytes can be eliminated if the whole measurement is performed in the cytoplasm-free cell extract; it allows for inspection if any interspin distances below 1.5 nm are present after folding of spin labeled oligonucleotides.

DEER measurements in the cell extract were performed in the way as close as possible to *in-cell* experiments. 1 μL of 4 mM stock solution of the spin labeled HT-repeat was added without mixing to 24 μL of oocyte cell extract with 20 % (v/v) of glycerol which was already placed in the 2 mm ID EPR tube. Incubation of the sample at room temperature served for diffusion and folding of the HT-repeat in the cytoplasmic environment of oocytes similar to the case inside cells. Consequently, the sample was shock-frozen in liquid nitrogen. The first and the shortest incubation time – the time delay between adding of the HT-repeat to the cell extract and freezing – was 0.5 min. The sample with 160 μM final concentration of the HT-repeat was measured with the conventional DEER setup and a pump π-pulse of 12 ns length. The lyophilization is not required for the samples in the cells extract and the sample handling becomes easier. This allows for performing a series of subsequent DEER measurement and thus allows for following conformational changes in the one single sample in solution. After the measurement at 0.5 min incubation time the sample was thawed to room temperature, incubated 11.5 min more and frozen again. Thus, at total incubation time of 12 min a new DEER measurement was carried out providing one more "snapshot" of interspin distance distributions which were formed during this new incubation period. One more thaw-freezing cycle with 8 min delay in between gave a total incubation time of 20 min at which the dipolar evolution curve was recorded again. At the first glance, the whole procedure may seem time- and effort-consuming but it provides unique access to observation of structural changes in solution while saving high amounts of an expensive sample. Time-resolution in this series of measurements is defined by complete freezing and thawing of the probe and is at least 20 s for total sample volume of 25 µL.

In order to get a full picture of folding of the HT-repeat in the cell extract the distance distribution at incubation time $t = 0$ min is required. It corresponds to the HT-repeat in the unfolded state as it is added to the cell extract. Thus, 1 µL of the HT-repeat was mixed with 24 µL of Tris-HCl buffer (pH 7.5) already containing 20 % (v/v) of glycerol. It was frozen in liquid nitrogen and measured. The HT-repeat in a salt-free buffer features a random coil conformation (CD spectra on Fig. 28a) with the most probable distance between nitroxides being 2.84 nm as estimated from Tikhonov regularization (Fig. 47).

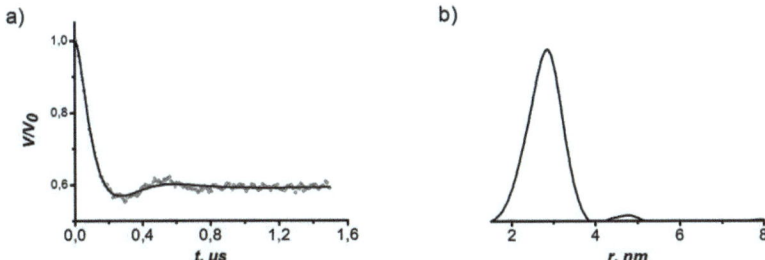

Figure 47. **DEER measurement on the HT-repeat in the salt-free buffer.** (a) Dipolar evolution curve (circles) with the fit from Tikhonov regularization (solid line) at α = 100 and (b) respective distance distribution.

Prior to DEER measurements in the cell extract the transversal relaxation times were determined for each particular incubation period. Its increase from 1.76 µs (0.5 min incubation time) up to 2.45 µs (20 min incubation time) is a result of a reduced local concentration due to diffusion of spin labeled HT-repeat molecules in the cell extract and reduction of nitroxides. Similar to *in-cell* DEER, each dipolar evolution curve for *in-extract* samples was recorded over 1.5 µs. The experimental dipolar evolution curves were corrected for intermolecular contributions using background functions derived from the same measurements of the single spin labeled HT-repeat. Owing to the short incubation time the dipolar evolution curve of 0.5-min-sample features a modulation depth of $\lambda = 0.42$ (Fig. 48a). The decrease of the modulation depth with increased incubation time (Fig. 48a) reflects lowering fraction of coupled spins due to reduction of nitroxide spin labels *in cellulo*. However, even after 20 min the $\lambda = 0.18$ allows for reliable extraction of interspin distances.

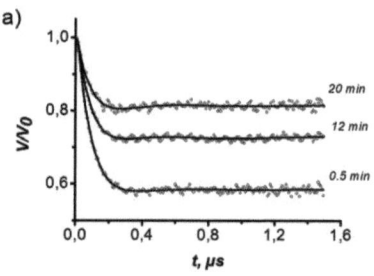

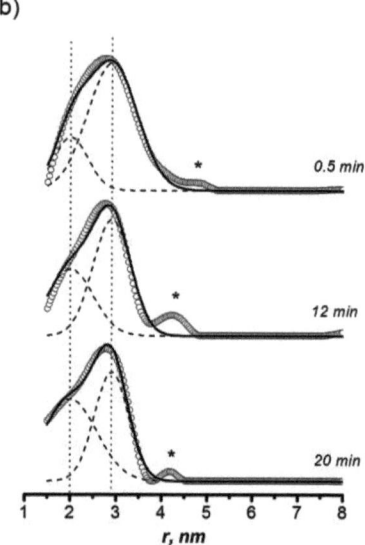

Figure 48. *In-extract* **DEER of the spin labeled HT-repeat.** (a) Experimental background-corrected dipolar evolution curves for different incubation times (circles) fitted with Tikhonov regularization (solid lines). (b) From Tikhonov regularization calculated distance distributions (circles), solid lines represent a superposition of two individual Gaussians (dashed) which maxima are fixed at 2.0 and 2.9 nm (vertical black dotted lines). Asterisks mark artifacts in distance distributions due to Tikhonov regularization.

The calculation of distances in the series of time-dependent DEER measurements is not a trivial task as one deals with conformational changes which are in progress and thus with a system in a non-equilibrium transition state. For this reason, model-free Tikhonov regularization was applied first to extract interspin distances. Already visual inspection of distance distributions shows clear changes from one incubation time to another (Fig. 48b, blue traces). Those distances provide only an envelope of distribution whereas no conclusions about G-quadruplex conformations in the sample can be met so far. If those distance distributions are fitted with two Gaussians where all five parameters (*i.e.* ratio between single Gaussians and their individual centers and widths) are varied, two maxima at 2.00 ± 0.05 nm and 2.90 ± 0.05 nm appear in all cases. These values correspond to the most probable interspin distances found inside cells (Table 6) and observations in cell extract seem to support *in-cell* findings. Thus, for the sake of better numerical characterization the distance distributions were fitted with fixed maxima at 2.0 nm and 2.9 nm. Those maxima are assigned to the parallel and the antiparallel structures, respectively. As dealing with transition from an unfolded state to a folded one, the variation of ratio and width of individual Gaussians allows for required flexibility of the fitting procedure with respect to conformational transitions of the HT-repeat.

For the two Gaussians with fixed maxima at 2.0 and 2.9 nm the increase of the fraction at 2.0 nm is observed in the distance distribution with longer incubation times (Fig. 48b; Table 6). That is, the HT-repeat adopts the parallel conformation which reaches approximately 50 % after 20 min. The peak at 2.9 nm cannot be solely assigned to the antiparallel-basket topology because it coincides with the most probable interspin distance in the unfolded HT-repeat (Fig. 47b). This also explains wider σ for 2.9 nm peak at 0.5 min incubation time which later gets narrower ($\sigma = 0.4$ nm). Only at incubation time of 20 min it can be assumed that the Gaussian at 2.9 nm corresponds to the antiparallel structure. Nitroxide spin labels in both quadruplexes are situated in the loop regions and point in the bulk solution and therefore are equally exposed to it. Consequently, nitroxides on both G-quadruplexes are assumed to be reduced at the same rate and the ratio between two conformations which was observed for λ = 0.18 (20 min incubation time) should be true for the whole system.

Table 6. Characteristic parameters for distance distributions obtained for the double spin labeled HT-repeat in cellular environment of *X. laevis* oocytes.

Experiment	Distance constraint r_1, nm	Width of distance distribution (HWHH) σ_1, nm	Distance constraint r_2, nm	Width of distance distribution (HWHH) σ_2, nm	Fraction of distance constraint r_1
In-extract (30 sec) [a]	2.0[b]	0.4	2.9[b]	0.6	21%
In-extract (12 min) [a]	2.0[b]	0.5	2.9[b]	0.4	37%
In-extract (20 min) [a]	2.0[b]	0.6	2.9[b]	0.4	49%
In-cell[c]	2.0 (±0.1)	0.7 (±0.1)	2.9 (±0.1)	0.4 (±0.1)	45% (±4%)

[a] distance distribution obtained by Tikhonov regularization and fitted with two Gaussians; [b] kept constant; [c] Two-Gauss-curve model.

Distances around below 1.5 nm cannot be detected accurately by DEER. The occurrence of short-range distances below 1.5 nm for the *in-extract* series (*e.g.* Fig. 47b) is therefore crosschecked by cw EPR studies in cell extract. Those allow for checking short-range distances by comparison of cw spectra from double spin labeled and single spin labeled samples at 120 K. The corresponding cw spectra showing subtle differences (Fig. 49b) support the distance distributions assumed for the *in-extract* series, while for the unfolded HT-repeat in salt-free buffer solution no distances below 1.5 nm were detected (Fig. 49a).

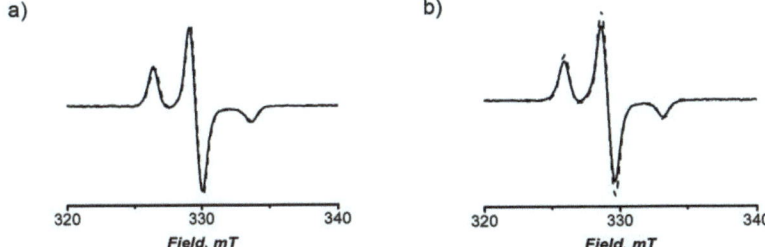

Figure 49. **Cw spectra at 120 K.** Single spin labeled (dashed line) and double spin labeled (solid line) HT-repeat in (a) the salt-free buffer and in (b) the cell extract after 12 min incubation time.

In summary, *in-cell* DEER was shown to be applicable to identify G-quadruplex topologies inside cells. The HT-repeat adopts a 1:1 mixture of the parallel-propeller and the antiparallel-basket conformations in *X. laevis* oocytes. The same result was observed for the HT-repeat in K^+-containing buffer and allowed for conclusion of the dominant role of K^+ ions over other environmental factors which might drive folding of the HT-repeat. DEER measurements in the cell extract support findings inside cells and also complement *in-cell* DEER by analysis of short-range distances and allow for monitoring of conformational changes in solution.

Conclusions

The current work deals with the development of EPR for biological applications. Based on existing *in vitro* techniques, *in-cell* approaches for determination of distance constraints by double electron-electron resonance (DEER) were established. After proof-of-principle experiments on rigid model systems, *in-cell* DEER has been applied to study formation of G-quadruplexes adopted by human telomeric DNA oligomers upon their microinjection into living cells.

Distances between two spin labels are derived from a DEER experiment. Their incorporation into DNA oligomers was achieved by inserting nitroxide-modified 2′-deoxyuridines at places of thymidines during solid-phase synthesis. By means of CD spectroscopy it was shown that nitroxides incorporated in this way in loop regions of G-quadruplexes do not disturb their initial structures. In spite of the inherent flexibility of loops, measured interspin distances are characteristic: they vary significantly for different G-quadruplex topologies providing a criterion for identification of those topologies. The most probable interspin distance can be predicted from existing high-resolution structural data and compared with reference measurements of known structures allowing for allocation to a certain structure. Even several conformations can be identified within a single experiment if the difference between the most probable interspin distances exceeds 0.3 nm. In the ion-dependent folding of the d[AGGG(TTAGGG)$_3$] the antiparallel-basket topology was found for Na$^+$ (in agreement with high-resolution NMR structure) and for the first time a 1:1 mixture of the antiparallel-basket and the parallel-propeller in the presence of K$^+$ ions was identified.

G-quadruplexes may alter folding topology if they are formed one after another within a long DNA sequence. This influence of neighboring G-quadruplexes was studied in a sequence with three G-quadruplex-forming blocks where one of them – in the middle or at the side – was selectively labeled with nitroxides. Consequent distance measurements allowed for conclusion on topologies adopted solely by the middle or the terminal part of the DNA sequence and thus allowed for building models for the folded human telomeric DNA in solution.

The intracellular environment is known to reduce nitroxides to EPR-silent hydroxylamines. Therefore, prior to novel *in-cell* distance measurements free nitroxides were deposited into cells via microinjection. Here, the reduction kinetics of two types of nitroxides was described within Michaelis-Menten formalism for enzymatic processes. Thus, an enzymatic origin of reduction *in cellulo* was proposed. It was found that inside *X. laevis* oocytes and in their extract the five member ring nitroxides (like PCA) are reduced much slower than their six member ring analogs. In principle, both six and five member ring nitroxides are suitable for *in-cell* DEER if no long incubation time is required. However, five member ring nitroxides are preferable due to longer half-life and thus easier handling.

Conformations of the d[AGGG(TTAGGG)$_3$] inside oocytes where nucleus-like conditions preside were studied after incubating the unfolded sequence for 15 min after injection. Two conformations at 1:1 ratio were found inside cells. These were the antiparallel-basket and the parallel-propeller. This result is the same as found in K$^+$-containing buffer. Thus, a dominant role of K$^+$ in *in-cell* folding of the human telomeric oligonucleotide was established. This result was also confirmed by *in-extract* measurement providing an alternative way to study conformations *in cellulo*. A time-dependent experiment allowed monitoring of the folding process.

DEER was shown to be applicable for structural identification goals of biological objects under their native conditions. *In-cell* DEER can provide means to analyze conformations and folding of nucleic acid structures inside living cells which has hitherto been unachievable.

List of references

[1] J. D. Watson, F. H. C. Crick, Molecular structure of nucleic acids - a structure for deoxyribose nucleic acid, *Nature* **1953**, *171*, 737-738.

[2] R. Chattopadhyaya, K. Grzeskowiak, R. E. Dickerson, Structure of a T4 hairpin loop on a Z-DNA stem and comparison with A-RNA and B-DNA loops, *J. Mol. Biol.* **1990**, *211*, 189-210.

[3] R. P. Goodman, I. A. T. Schaap, C. F. Tardin, C. M. Erben, R. M. Berry, C. F. Schmidt, A. J. Turberfield, Rapid chiral assembly of rigid DNA building blocks for molecular nanofabrication, *Science* **2005**, *310*, 1661-1665.

[4] M. Gellert, M. N. Lipsett, D. R. Davies, Helix formation by guanylic acid, *Proc. Natl. Acad. Sci. U. S. A.* **1962**, *48*, 2013-2018.

[5] M. A. Keniry, Quadruplex structures in nucleic acids, *Biopolymers* **2001**, *56*, 123-146.

[6] P. Hazel, J. Huppert, S. Balasubramanian, S. Neidle, Loop-length-dependent folding of G-quadruplexes, *J. Am. Chem. Soc.* **2004**, *126*, 16405-16415.

[7] E. H. Blackburn, Structure and function of telomeres, *Nature* **1991**, *350*, 569-573.

[8] G. B. Morin, The human telomere terminal transferase enzyme is a ribonucleoprotein that synthesizes TTAGGG repeats, *Cell* **1989**, *59*, 521-529.

[9] W. E. Wright, V. M. Tesmer, K. E. Huffman, S. D. Levene, J. W. Shay, Normal human chromosomes have long G-rich telomeric overhangs at one end, *Genes Dev.* **1997**, *11*, 2801-2809.

[10] K. E. Huffman, S. D. Levene, V. M. Tesmer, J. W. Shay, W. E. Wright, Telomere shortening is proportional to the size of the G-rich telomeric 3'-overhang, *J. Biol. Chem.* **2000**, *275*, 19719-19722.

[11] J. L. Mergny, C. Helene, G-quadruplex DNA: A target for drug design, *Nat. Med.* **1998**, *4*, 1366-1367.

[12] A. M. Olovnikov, Theory of marginotomy - incomplete copying of template margin in enzymic-synthesis of polynucleotides and biological significance of phenomenon, *J. Theor. Biol.* **1973**, *41*, 181-190.

[13] C. B. Harley, A. B. Futcher, C. W. Greider, Telomeres shorten during aging of human fibroblasts, *Nature* **1990**, *345*, 458-460.

[14] A. J. Zaug, E. R. Podell, T. R. Cech, Human POT1 disrupts telomeric G-quadruplexes allowing telomerase extension in vitro, *Proc. Natl. Acad. Sci. U. S. A.* **2005**, *102*, 10864-10869.

[15] D. Y. Sun, B. Thompson, B. E. Cathers, M. Salazar, S. M. Kerwin, J. O. Trent, T. C. Jenkins, S. Neidle, L. H. Hurley, Inhibition of human telomerase by a G-quadruplex-interactive compound, *J. Med. Chem.* **1997**, *40*, 2113-2116.

[16] K. C. Healy, Telomeric dynamics and telomerase activation in tumor progression - prospects for prognosis and therapy, *Oncol. Res.* **1995**, *7*, 121-130.

[17] J. L. Huppert, S. Balasubramanian, Prevalence of quadruplexes in the human genome, *Nucleic Acids Res.* **2005**, *33*, 2908-2916.

[18] A. K. Todd, M. Johnston, S. Neidle, Highly prevalent putative quadruplex sequence motifs in human DNA, *Nucleic Acids Res.* **2005**, *33*, 2901-2907.

[19] Y. Wang, D. J. Patel, Solution structure of the human telomeric repeat d[AG(3)(T(2)AG(3))3] G-tetraplex, *Structure* **1993**, *1*, 263-282.

[20] G. N. Parkinson, M. P. H. Lee, S. Neidle, Crystal structure of parallel quadruplexes from human telomeric DNA, *Nature* **2002**, *417*, 876-880.

[21] K. N. Luu, A. T. Phan, V. Kuryavyi, L. Lacroix, D. J. Patel, Structure of the human telomere in K(+) solution: An intramolecular (3+1) G-quadruplex scaffold, *J. Am. Chem. Soc.* **2006**, *128*, 9963-9970.

[22] K. W. Lim, S. Amrane, S. Bouaziz, W. Xu, Y. Mu, D. J. Patel, K. N. Luu, A. T. Phan, Structure of the human telomere in K(+) solution: a stable

basket-type G-quadruplex with only two G-tetrad layers, *J. Am. Chem. Soc.* **2009**, *131*, 4301-4309.

[23] A. Ambrus, D. Chen, J. X. Dai, T. Bialis, R. A. Jones, D. Z. Yang, Human telomeric sequence forms a hybrid-type intramolecular G-quadruplex structure with mixed parallel/antiparallel strands in potassium solution, *Nucleic Acids Res.* **2006**, *34*, 2723-2735.

[24] J. Li, J. J. Correia, L. Wang, J. O. Trent, J. B. Chaires, Not so crystal clear: the structure of the human telomere G-quadruplex in solution differs from that present in a crystal, *Nucleic Acids Res.* **2005**, *33*, 4649-4659.

[25] B. Heddi, P. Anh Tuan, Structure of human telomeric DNA in crowded solution, *J. Am. Chem. Soc.* **2011**, *133*, 9824-9833.

[26] N. H. Campbell, G. N. Parkinson, Crystallographic studies of quadruplex nucleic acids, *Methods* **2007**, *43*, 252-263.

[27] M. Webba da Silva, NMR methods for studying quadruplex nucleic acids, *Methods* **2007**, *43*, 264-277.

[28] R. Hänsel, S. Foldynova-Trantirkova, F. Löhr, J. Buck, E. Bongartz, E. Bamberg, H. Schwalbe, V. Dötsch, L. Trantirek, Evaluation of parameters critical for observing nucleic acids inside living Xenopus laevis oocytes by in-cell NMR spectroscopy, *J. Am. Chem. Soc.* **2009**, *131*, 15761-15768.

[29] S. Paramasivan, I. Rujan, P. H. Bolton, Circular dichroism of quadruplex DNAs: Applications to structure, cation effects and ligand binding, *Methods* **2007**, *43*, 324-331.

[30] J. Kypr, I. Kejnovska, D. Renciuk, M. Vorlickova, Circular dichroism and conformational polymorphism of DNA, *Nucleic Acids Res.* **2009**, *37*, 1713-1725.

[31] S. Masiero, R. Trotta, S. Pieraccini, S. De Tito, R. Perone, A. Randazzo, G. P. Spada, A non-empirical chromophoric interpretation of CD spectra of DNA G-quadruplex structures, *Org. Biomol. Chem.* **2010**, *8*, 2683-2692.

[32] A. Benz, V. Singh, T. U. Mayer, J. S. Hartig, Identification of novel quadruplex ligands from small molecule libraries by FRET-based high-throughput screening, *ChemBioChem* **2011**, *12*, 1422-1426.

[33] T. Simonsson, R. Sjoback, DNA tetraplex formation studied with fluorescence resonance energy transfer, *J. Biol. Chem.* **1999**, *274*, 17379-17383.

[34] T. I. Gaynutdinov, R. D. Neumann, I. G. Panyutin, Structural polymorphism of intramolecular quadruplex of human telomeric DNA: effect of cations, quadruplex-binding drugs and flanking sequences, *Nucleic Acids Res.* **2008**, *36*, 4079-4087.

[35] T. Miura, G. J. Thomas, Structural polymorphism of telomere DNA - interquadruplex and duplex-quadruplex conversions probed by Raman-spectroscopy, *Biochemistry* **1994**, *33*, 7848-7856.

[36] N. Smargiasso, F. Rosu, W. Hsia, P. Colson, E. S. Baker, M. T. Bowers, E. De Pauw, V. Gabelica, G-quadruplex DNA assemblies: Loop length, cation identity, and multimer formation, *J. Am. Chem. Soc.* **2008**, *130*, 10208-10216.

[37] J. E. Redman, Surface plasmon resonance for probing quadruplex folding and interactions with proteins and small molecules, *Methods* **2007**, *43*, 302-312.

[38] I. A. Pedroso, L. F. Duarte, G. Yanez, K. Burkewitz, T. M. Fletcher, Sequence specificity of inter- and intramolecular G-quadruplex formation by human telomeric DNA, *Biopolymers* **2007**, *87*, 74-84.

[39] P. B. Ayscough, *Electron spin resonance in chemistry*, Methuen, London, **1967**.

[40] N. M. Atherton, *Electron Spin Resonance. Theory and applications*, Horwood, Chichester, **1973**.

[41] D. J. E. Ingram, *Biological and biochemical application of electron spin resonance*, Hilger, London, **1969**.

[42] A. Abragam, Bleaney, B., *Electron paramagnetic resonance of transition ions*, Clarendon Press, Oxford, **1970**.

[43] A. Schweiger, Jeschke, G., *Principles of pulse electron paramagnetic resonance*, Oxford Univ. Press, Oxford, **2005**.

[44] O. Schiemann, P. Cekan, D. Margraf, T. F. Prisner, S. T. Sigurdsson, Relative orientation of rigid nitroxides by PELDOR: Beyond distance measurements in nucleic acids, *Angew. Chem. Int. Ed.* **2009**, *48*, 3292-3295.

[45] E. L. Hahn, Spin Echoes, *Phys. Rev.* **1950**, *80*, 580-594.

[46] J. E. Banham, C. M. Baker, S. Ceola, I. J. Day, G. H. Grant, E. J. J. Groenen, C. T. Rodgers, G. Jeschke, C. R. Timmel, Distance measurements in the borderline region of applicability of CW EPR and DEER: A model study on a homologous series of spin-labelled peptides, *J. Magn. Reson.* **2008**, *191*, 202-218.

[47] A. D. Milov, A. B. Ponomarev, Y. D. Tsvetkov, Electron electron double-resonance in electron-spin echo - model biradical systems and the sensitized photolysis of decalin, *Chem. Phys. Lett.* **1984**, *110*, 67-72.

[48] G. Jeschke, Distance measurements in the nanometer range by pulse EPR, *ChemPhysChem* **2002**, *3*, 927-932.

[49] A. M. Raitsimring, K. M. Salikhov, B. A. Umanskii, Y. D. Tsvetkov, Instantaneous diffusion in electron-spin echo of paramagnetic centers stabilized in solid matrices, *Fiz. Tverd. Tela* **1974**, *16*, 756-766.

[50] M. Pannier, S. Veit, A. Godt, G. Jeschke, H. W. Spiess, Dead-time free measurement of dipole-dipole interactions between electron spins, *J. Magn. Reson.* **2000**, *142*, 331-340.

[51] G. Jeschke, Y. Polyhach, Distance measurements on spin-labelled biomacromolecules by pulsed electron paramagnetic resonance, *Phys. Chem. Chem. Phys.* **2007**, *9*, 1895-1910.

[52] L. J. Berliner, J. Reuben, Spin labeling. Theory and Applications in *Biological Magnetic Resonance, Vol. 8*, Plenum Press, New York, **1989**, p. 650.

[53] J. L. Holtzman, Spin labeling in pharmacology, Academic Press, London, **1984**, p. 229.

[54] J. F. W. Keana, F. L. van Nice, Influence of structure of the reduction of nitroxide MRI contrast-enhancing agents by ascorbate, *Physiol. Chem. Phys. Med. NMR* **1984**, *16*, 477-480.

[55] A. R. Forrester, S. P. Hepburn, Nitroxide radicals. Part XV. p-methoxyphenyl and p-phenoxy-phenyl t-butyl nitroxides, *J. Chem. Soc. Perkin Trans. 1* **1974**, 2208-2213.

[56] R. Ramasseul, A. Rassat, Nitroxides. XXXIII. Radicaux: nitroxydes pyrroliques encombres. Un pyrryloxyle stable, *Bull. Soc. Chim. Fr.* **1970**, 4330-4341.

[57] C. Berti, M. Colonna, L. Greci, L. Marchetti, Stable nitroxide radicals from acridine N-oxides with Grignard-reagents, *Gazz. Chim. Ital.* **1978**, *108*, 659-664.

[58] J. F. W. Keana, Synthesis and chemistry of nitroxide spin labels in *Spin labeling in pharmacology* (Ed.: J. L. Holtzman), Academic Press, London, **1984**, pp. 2-86.

[59] D. F. Bowman, T. Gillan, K. U. Ingold, Kinetic applications of electron paramagnetic resonance spectroscopy. 3. Self-ractions of dialkyl nitroxide radicals, *J. Am. Chem. Soc.* **1971**, *93*, 6555-6561.

[60] P. Jost, L. J. Libertini, V. C. Hebert, O. H. Griffith, Lipid spin labels in lecithin multilayers - study of motion along fatty acid chains, *J. Mol. Biol.* **1971**, *59*, 77-98.

[61] G. Jeschke, Kurze Einführung in die electronenparamagnetische Resonanzspektroskopie, Vorlesungsskript ed., Konstanz, **2008**, p. 103.

[62] K. Möbius, A. Savitsky, C. Wegener, M. Rato, M. Fuchs, A. Schnegg, A. A. Dubinskii, Y. A. Grishin, I. A. Grigor'ev, M. Kuhn, D. Duche, H. Zimmermann, H. J. Steinhoff, Combining high-field EPR with site-directed spin labeling reveals unique information on proteins in action, *Magn. Reson. Chem.* **2005**, *43*, S4-S19.

[63] S. Stoll, A. Schweiger, EasySpin, a comprehensive software package for spectral simulation and analysis in EPR, *J. Magn. Reson.* **2006**, *178*, 42-55.

[64] A. G. Redfield, The theory of relaxation processes, *Adv. Magn. Reson.* **1965**, *1*, 1-32.

[65] J. H. Freed, Theory of slow tumbling ESR spectra for nitroxides in *Spin labeling: Theory and Applications, Vol. 1* (Ed.: L. J. Berliner), Academic Press, New York, **1976**, pp. 53-132.

[66] C. Beier, H.-J. Steinhoff, A structure-based simulation approach for electron paramagnetic resonance spectra using molecular and stochastic dynamics simulations, *Biophys. J.* **2006**, *91*, 2647-2664.

[67] W. L. Hubbell, D. S. Cafiso, C. Altenbach, Identifying conformational changes with site-directed spin labeling, *Nat. Struct. Biol.* **2000**, *7*, 735-739.

[68] O. Schiemann, N. Piton, J. Plackmeyer, B. E. Bode, T. F. Prisner, J. W. Engels, Spin labeling of oligonucleotides with the nitroxide TPA and use of PELDOR, a pulse EPR method, to measure intramolecular distances, *Nat. Protoc.* **2007**, *2*, 904-923.

[69] G. Sicoli, F. Wachowius, M. Bennati, C. Hoebartner, Probing secondary structures of spin-labeled RNA by pulsed EPR spectroscopy, *Angew. Chem. Int. Ed.* **2010**, *49*, 6443-6447.

[70] C. Giordano, F. Pedone, P. Fattibene, L. Cellai, Oligonucleotide labeling: Synthesis of a new spin-labeled 2'-deoxyguanosine analogue, *Nucleosides, Nucleotides Nucleic Acids* **2000**, *19*, 1301-1310.

[71] N. Piton, Y. Mu, G. Stock, T. F. Prisner, O. Schiemann, J. W. Engels, Base-specific spin-labeling of RNA for structure determination, *Nucleic Acids Res.* **2007**, *35*, 3128-3143.

[72] P. Cekan, S. T. Sigurdsson, Identification of single-base mismatches in duplex DNA by EPR spectroscopy, *J. Am. Chem. Soc.* **2009**, *131*, 18054-18056.

[73] N. Barhate, P. Cekan, A. P. Massey, S. T. Sigurdsson, A nucleoside that contains a rigid nitroxide spin label: A fluorophore in disguise, *Angew. Chem. Int. Ed.* **2007**, *46*, 2655-2658.

[74] C. R. Toppin, G. T. Pauly, P. Devanesan, D. Kryak, A. M. Bobst, 3 novel spin-labeled substrates for enzymatic incorporation into nucleic-acid lattices, *Helv. Chim. Acta* **1986**, *69*, 345-349.

[75] E. J. Hustedt, J. J. Kirchner, A. Spaltenstein, P. B. Hopkins, B. H. Robinson, Monitoring DNA dynamics using spin-labels with different independent mobilities, *Biochemistry* **1995**, *34*, 4369-4375.

[76] B. H. Robinson, C. Mailer, G. Drobny, Site-specific dynamics in DNA: Experiments, *Annu. Rev. Biophys. Biomol. Struct.* **1997**, *26*, 629-658.

[77] T. Okonogi, A. W. Reese, S. C. Alley, P. B. Hopkins, B. H. Robinson, Flexibility of duplex DNA on the submicrosecond timescale, *Biophys. J.* **1999**, *77*, 3256-3276.

[78] N. D. Sinha, J. Biernat, J. McManus, H. Koster, Polymer support oligonucleotide synthesis. 18. Use of beta-cyanoethyl-N,N-dialkylamino-N-morpholino phosphoramidite of deoxynucleosides for the synthesis of DNA fragments simplifying deprotection and isolation of the final product, *Nucleic Acids Res.* **1984**, *12*, 4539-4557.

[79] S. Obeid, M. Yulikov, G. Jeschke, A. Marx, Enzymatic synthesis of multiple spin-labeled DNA, *Angew. Chem. Int. Ed.* **2008**, *47*, 6782-6785.

[80] U. Jakobsen, S. A. Shelke, S. Vogel, S. T. Sigurdsson, Site-directed spin-labeling of nucleic acids by click chemistry: detection of abasic sites in duplex DNA by EPR spectroscopy, *J. Am. Chem. Soc.* **2010**, *132*, 10424-10428.

[81] Q. Cai, A. K. Kusnetzow, W. L. Hubbell, I. S. Haworth, G. P. C. Gacho, N. Van Eps, K. Hideg, E. J. Chambers, P. Z. Qin, Site-directed spin labeling measurements of nanometer distances in nucleic acids using a sequence-independent nitroxide probe, *Nucleic Acids Res.* **2006**, *34*, 4722-4730.

[82] S. Nagahara, A. Murakami, K. Makino, Spin-labeled oligonucleotides site specifically labeled at the internucleotide linkage - separation of

stereoisomeric probes and EPR spectroscopical detection of hybrid formation in solution, *Nucleosides, Nucleotides* **1992**, *11*, 889-901.

[83] R. Ward, D. J. Keeble, H. El-Mkami, D. G. Norman, Distance determination in heterogeneous DNA model systems by pulsed EPR, *ChemBioChem* **2007**, *8*, 1957-1964.

[84] O. Schiemann, N. Piton, Y. G. Mu, G. Stock, J. W. Engels, T. F. Prisner, A PELDOR-based nanometer distance ruler for oligonucleotides, *J. Am. Chem. Soc.* **2004**, *126*, 5722-5729.

[85] A. Marko, V. Denysenkov, D. Margraf, P. Cekan, O. Schiemann, S. T. Sigurdsson, T. F. Prisner, Conformational flexibility of DNA, *J. Am. Chem. Soc.* **2011**, *133*, 13375-13379.

[86] C. Policar, J. B. Waern, M.-A. Plamont, S. Clede, C. Mayet, R. Prazeres, J.-M. Ortega, A. Vessieres, A. Dazzi, Subcellular IR imaging of a metal-carbonyl moiety using photothermally induced resonance, *Angew. Chem. Int. Ed.* **2011**, *50*, 860-864.

[87] W. Xie, L. Su, A. Shen, A. Materny, J. Hu, Application of surface-enhanced Raman scattering in cell analysis, *J. Raman Spectrosc.* **2011**, *42*, 1248-1254.

[88] C. Brackmann, A. Bengtsson, M. L. Alminger, U. Svanberg, A. Enejder, Visualization of beta-carotene and starch granules in plant cells using CARS and SHG microscopy, *J. Raman Spectrosc.* **2011**, *42*, 586-592.

[89] H. Chen, S. Kim, L. Li, S. Wang, K. Park, J.-X. Cheng, Release of hydrophobic molecules from polymer micelles into cell membranes revealed by Förster resonance energy transfer imaging, *Proc. Natl. Acad. Sci. U. S. A.* **2008**, *105*, 6596-6601.

[90] L. J. Berliner, H. Fujii, Magnetic-resonance imaging of biological specimens by electron-paramagnetic resonance of nitroxide spin labels, *Science* **1985**, *227*, 517-519.

[91] H. M. Swartz, K. Chen, M. Pals, M. Sentjurc, P. D. Morse, Hypoxia-sensitive NMR contrast agents, *Magn. Reson. Med.* **1986**, *3*, 169-174.

[92] G. Kohler, C. Milstein, Continuous cultures of fused cells secreting antibody of predefined specificity, *Nature* **1975**, *256*, 495-497.

[93] T. A. Kunkel, Rapid and efficient site-specific mutagenesis without phenotypic selection, *Proc. Natl. Acad. Sci. U. S. A.* **1985**, *82*, 488-492.

[94] H. M. Swartz, M. Sentjurc, P. D. Morse, Cellular-metabolism of water-soluble nitroxides - effect on rate of reduction of cell nitroxide ratio, oxygen concentrations and permeability of nitroxides, *Biochim. Biophys. Acta* **1986**, *888*, 82-90.

[95] S. Au, S. Cohen, N. Pante, Microinjection of Xenopus laevis oocytes as a system for studying nuclear transport of viruses, *Methods* **2010**, *51*, 114-120.

[96] S. P. Williams, P. M. Haggie, K. M. Brindle, F-19 NMR measurements of the rotational mobility of proteins in vivo, *Biophys. J.* **1997**, *72*, 490-498.

[97] Z. Serber, V. Dötsch, In-cell NMR spectroscopy, *Biochemistry* **2001**, *40*, 14317-14323.

[98] H. P. Hu, G. Sosnovsky, H. M. Swartz, Simultaneous measurements of the intracellular and extracellular oxygen concentration in viable cells, *Biochim. Biophys. Acta* **1992**, *1112*, 161-166.

[99] E. J. Rauckman, G. M. Rosen, J. Cavagnaro, Norcocaine nitroxide - a potential hepatotoxic metabolite of cocaine, *Mol. Pharmacol.* **1982**, *21*, 458-463.

[100] Y. Y. Woldman, S. V. Semenov, A. A. Bobko, I. A. Kirilyuk, J. F. Polienko, M. A. Voinov, E. G. Bagryanskaya, V. V. Khramtsov, Design of liposome-based pH sensitive nanoSPIN probes: nano-sized particles with incorporated nitroxides, *Analyst* **2009**, *134*, 904-910.

[101] A. Feldman, E. Wildman, G. Bartolinini, L. H. Piette, In vivo electron-spin resonance in rats, *Phys. Med. Biol.* **1975**, *20*, 602-612.

[102] E. J. Rauckman, Rosen, G. M., Griffeth, L. K., Enzymatic reactions of spin labels in *Spin labeling in pharmacology* (Ed.: J. L. Holtzman), Academic Press, London, **1984**, pp. 175-192.

[103] N. Kocherginsky, H. M. Swartz, *Nitroxide spin labels: Reactions in biology and chemistry*, CRC Press, Boca Raton, **1995**.

[104] A. A. Bobko, I. A. Kirilyuk, I. A. Grigor'ev, J. L. Zweier, V. V. Khramtsov, Reversible reduction of nitroxides to hydroxylamines: Roles for ascorbate and glutathione, *Free Radical Bio. Med.* **2007**, *42*, 404-412.

[105] E. J. Rauckman, G. M. Rosen, B. B. Kitchell, Superoxide radical as an intermediate in the oxidation of hydroxylamines by mixed-function amine oxidase, *Mol. Pharmacol.* **1979**, *15*, 131-137.

[106] D. A. Butterfield, A. L. Crumbliss, D. B. Chestnut, Radical decay kinetics in ferrocytochrome c model membranes - spin label study, *J. Am. Chem. Soc.* **1975**, *97*, 1388-1393.

[107] G. M. Rosen, E. J. Rauckman, Formation and reduction of a nitroxide radical by liver-microsomes, *Biochem. Pharmacol.* **1977**, *26*, 675-678.

[108] A. Stier, I. Reitz, Radical production in amine oxidation by liver microsomes, *Xenobiotica* **1971**, *1*, 499-500.

[109] G. M. Rosen, E. J. Rauckman, K. W. Hanck, Selective bioreduction of nitroxides by rat-liver microsomes, *Toxicol. Lett.* **1977**, *1*, 71-74.

[110] R. Igarashi, T. Sakai, H. Hara, T. Tenno, T. Tanaka, H. Tochio, M. Shirakawa, Distance determination in proteins inside Xenopus laevis oocytes by double electron-electron resonance experiments, *J. Am. Chem. Soc.* **2010**, *132*, 8228-8229.

[111] P. M. Gannett, E. Darian, J. H. Powell, E. M. Johnson, A short procedure for synthesis of 4-ethynyl-2,2,6,6-tetramethyl-3,4-dehydro-piperidine-1-oxyl nitroxide, *Synth. Commun.* **2001**, *31*, 2137-2141.

[112] A. Spaltenstein, B. H. Robinson, P. B. Hopkins, Sequence-dependent and structure-dependent DNA-base dynamics - Synthesis, structure, and dynamics of site and sequence specifically spin-labeled DNA, *Biochemistry* **1989**, *28*, 9484-9495.

[113] S. W. Stork, M. W. Makinen, Facile synthesis of 3-formyl-2,2,5,5-tetramethyl-1-oxopyrroline, *Synthesis* **1999**, 1309-1312.

[114] J. J. Eppig, M. L. Steckman, Comparison of exogenous energy-sources for in vitro maintenance of follicle cell-free Xenopus-laevis oocytes, *In Vitro* **1976**, *12*, 173-179.

[115] J. B. Gurdon, Injected nuclein in frog oocytes - fate, enlargement, and chromatin dispersal, *J. Embryol. Exp. Morph.* **1976**, *36*, 523-540.

[116] J. N. Dumont, Oogenesis in Xenopus-laevis (daudin). 1. Stages of oocyte development in laboratory maintained animals, *J. Morphol.* **1972**, *136*, 153-179.

[117] A. W. Murray, Cell-cycle extracts, *Method. Cell. Biol.* **1991**, *36*, 581-605.

[118] A. Desai, A. Murray, T. J. Mitchison, C. E. Walczak, The use of Xenopus egg extracts to study mitotic spindle assembly and function in vitro, *Method. Cell. Biol.* **1999**, *61*, 385-412.

[119] G. Jeschke, V. Chechik, P. Ionita, A. Godt, H. Zimmermann, J. Banham, C. R. Timmel, D. Hilger, H. Jung, DeerAnalysis2006 - a comprehensive software package for analyzing pulsed ELDOR data, *Appl. Magn. Reson.* **2006**, *30*, 473-498.

[120] Y. W. Chiang, P. P. Borbat, J. H. Freed, The determination of pair distance distributions by pulsed ESR using Tikhonov regularization, *J. Magn. Reson.* **2005**, *172*, 279-295.

[121] P. Z. Qin, S. E. Butcher, J. Feigon, W. L. Hubbell, Quantitative analysis of the isolated GAAA tetraloop/receptor interaction in solution: A site-directed spin labeling study, *Biochemistry* **2001**, *40*, 6929-6936.

[122] A. Spaltenstein, B. H. Robinson, P. B. Hopkins, A rigid and nonperturbing probe for duplex DNA motion, *J. Am. Chem. Soc.* **1988**, *110*, 1299-1301.

[123] A. Okamoto, T. Inasaki, I. Saito, Nitroxide-labeled guanine as an ESR spin probe for structural study of DNA, *Bioorg. Med. Chem. Lett.* **2004**, *14*, 3415-3418.

[124] J. T. Edward, Molecular volumes and the Stokes-Einstein equation, *J. Chem. Educ.* **1970**, *47*, 261-270.

[125] G. Jeschke, Pannier, M., Spiess, H. W., Double electron-electron resonance in *Distance measurements in biological systems by EPR, Vol. 19* (Ed.: L. J. Berliner, Eaton G. R., Eaton S. S.), Kluwer Academics, New York, **2000**, pp. 493-512.

[126] G. Bacic, M. J. Nilges, R. L. Magin, T. Walczak, H. M. Swartz, In vivo localized ESR spectroscopy reflecting metabolism, *Magn. Reson. Med.* **1989**, *10*, 266-272.

[127] L. Michaelis, M. L. Menten, Die Kinetik der Intvertinwirkung, *Biochem. Z.* **1913**, *49*, 333-369.

[128] M. A. Savageau, Enzyme kinetics in vitro and in vivo: Michaelis-Menten revisited, *Prin. Med. Biol.* **1995**, *4*, 93-146.

[129] I. Krstic, R. Hänsel, O. Romainczyk, J. W. Engels, V. Dötsch, T. F. Prisner, Long-range distance measurements on nucleic acids in cells by pulsed EPR spectroscopy, *Angew. Chem. Int. Ed.* **2011**, *50*, 5070-5074.

Additional content

List of abbreviations

bp	base pair
CARS	coherent anti-Stokes Raman scattering
CD	circular dichroism
CPG	controlled porous glass
CSF	cryostatic factor arrested
cw	continuous wave
DEER	double electron-electron resonance
DNA	deoxyribonucleic acid
DTBN	di-*tert*-butylnitroxide
DTT	dithiothreitol
EDTA	ethylenediaminetetraacetic acid
EPR	electron paramagnetic resonance
FAD	flavin adenine dinucleotide
FAM	carboxyfluorescein
FRET	Förster resonance energy transfer
FT	Fourier transformation
HPLC	high pressure liquid chromatography
HWHH	half width at half height
ID	inner diameter
IR	infrared
MD	molecular dynamics
MRI	magnetic resonance imaging

MTSL	(1-oxyl-2,2,5,5-tetramethylpyrroline-3-methyl)-methanethiosulfonate
$NADP^+$	nicotinamide adenine dinucleotide phosphate
NADPH	nicotinamide adenine dinucleotide phosphate (reduced form)
NMR	nuclear magnetic resonance
nt	nucleotide
OD	outer diameter
PAGE	polyacrylamide gel electrophoresis
PEG	polyethyleneglycol
PELDOR	pulsed electron double resonance
PCA	3-carboxy-2,2,5,5-tetramethylpyrrolidinyl-1-oxy
r.m.s.d.	root mean square deviation
RNA	ribonucleic acid
SDSL	site-directed spin labelling
SEM	standard error of mean
SOD	superoxide dismutase
SPR	surface plasmon resonance
TAMRA	tetramethylrodamine
TEMPA	2,2,6,6-tetramethyl-3,4-dehydropiperidine-N-oxyl-4-acetylene
TEMPO	2,2,6,6-tetramethyl-N-oxyl
TOAC	2,2,6,6-tetramethylpiperidine-N-oxyl-4-amino-4-carboxylic acid
TPA	2,2,5,5-tetramethylpyrroline-1-oxyl-3-acetylene
X. laevis	*Xenopus laevis*

Supporting material

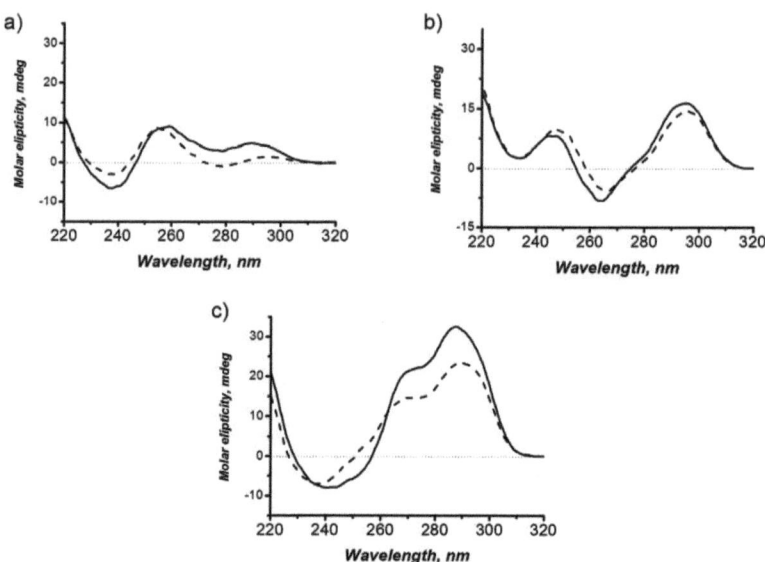

Figure A1. **CD spectra of the HT-control.** Unlabeled (solid line) and double spin labeled (dashed line) oligonucleotide in (a) a salt-free buffer, (b) Na^+-containing buffer and (c) K^+-containing buffer.

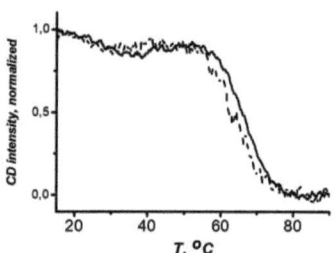

Figure A2. **Thermal denaturation profile for the HT-repeat in K^+-containing buffer detected by CD at 273 nm.** Unlabeled and double spin labeled oligonucleotides are presented in solid and dashed lines, respectively.

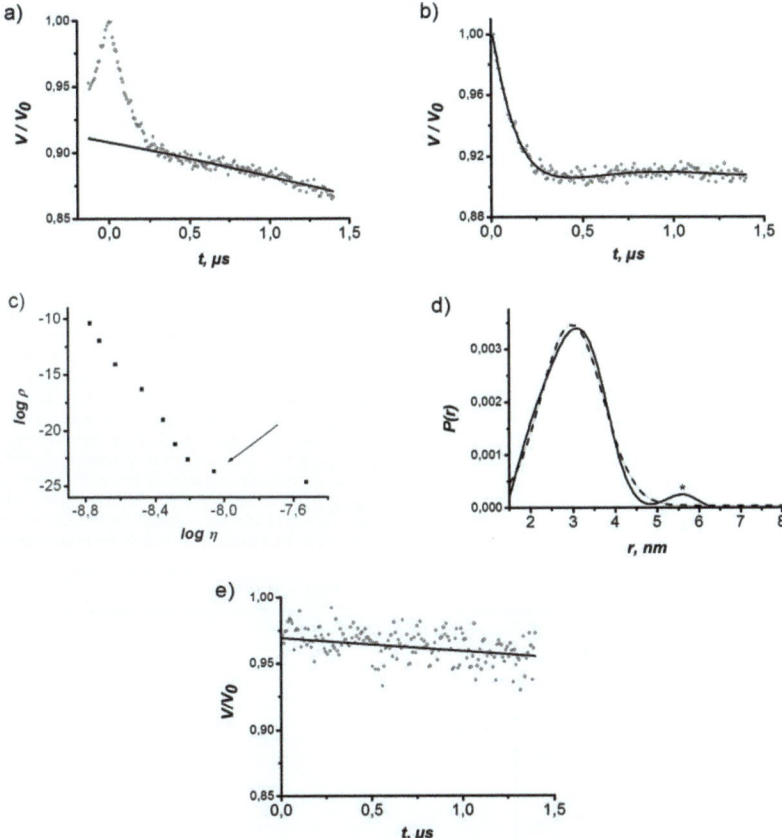

Figure A3. Analysis of the dipolar evolution curve measured for the spin labeled HT-repeat in Na$^+$-containing buffer. (a) Dipolar evolution curve prior to background correction (circles) with background function (solid line). (b) Background corrected dipolar evolution curve (circles) with fit from Tikhonov regularization at α = 10000. (c) The L-curve, optimal alpha is indicated by an arrow. (d) Distance distribution resulted from the Tikhonov regularization at α = 10000 (solid line) and Gaussian fit (dashed line) of this distribution. The asterisk marks an artefact due to Tikhonov regularization. (d) Dipolar evolution curve measured for the single spin labeled HT-repeat (circles) with the fit corresponding to the three-dimensional distribution (solid line).

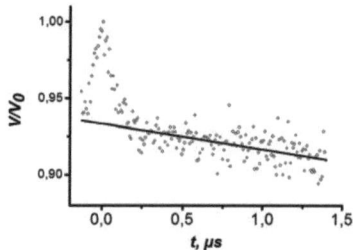

Figure A4. Dipolar evolution curves measured for the HT-control in K⁺-containing buffer. The dipolar evolution curve prior to background correction is shown in circles with the background function shown as solid line.

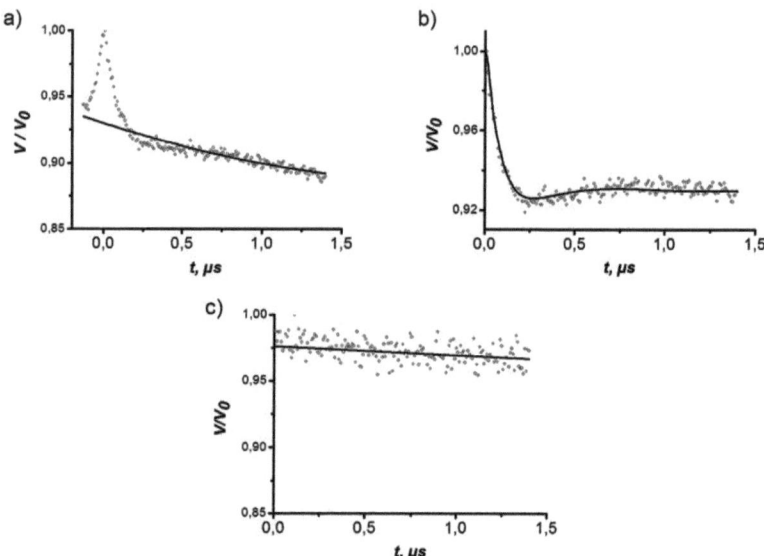

Figure A5. DEER on the double spin labeled HT-repeat in K⁺-containing buffer. (a) Dipolar evolution curve prior to background correction (circles) with the background function (solid line). (b) Dipolar evolution curve fitted with Tikhonov regularization. (c) Dipolar evolution curve measured for the single spin labeled HT-repeat (circles) with the fit corresponding to the three-dimensional distribution (solid line).

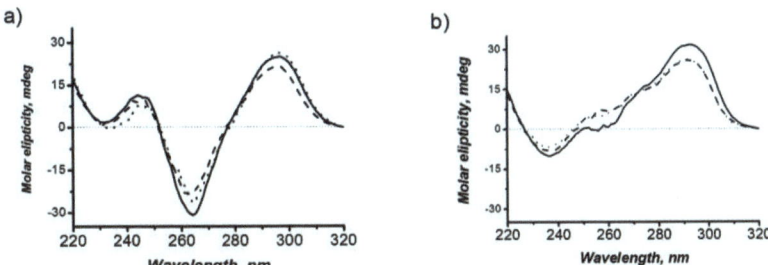

Figure A6. CD spectra of LongHT oligonucleotides. Unlabeled (solid line), single spin labeled at 5-position (dashed line), and double spin labeled at positions 29 and 35 (dotted line) LongHT in (a) Na^+-containing and (b) K^+-containing buffers.

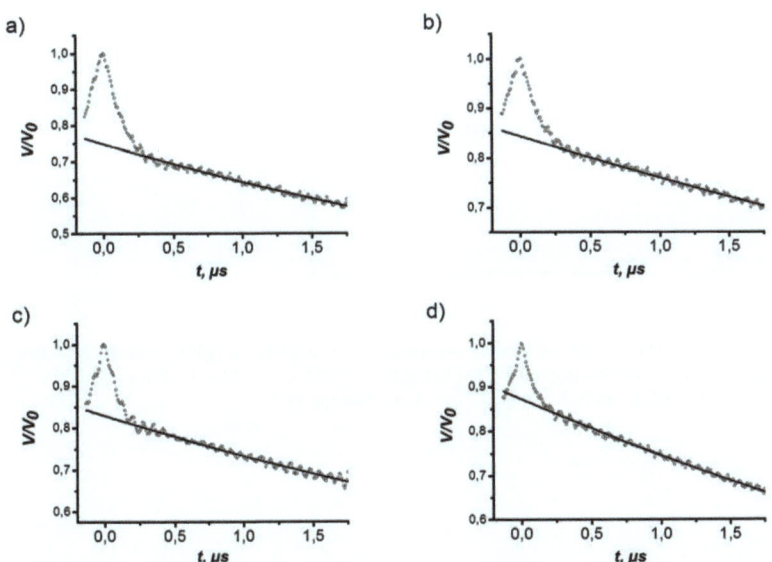

Figure A7. Dipolar evolution curves prior to background correction (circles) with background functions (solid line) for double spin labeled LongHT oligonucleotides. (a) LongHT^{side} in Na^+-containing buffer, (b) LongHT^{mid} in Na^+-containing buffer, (c) LongHT^{side} in K^+-containing buffer, and (d) LongHT^{mid} in K^+-containing buffer.

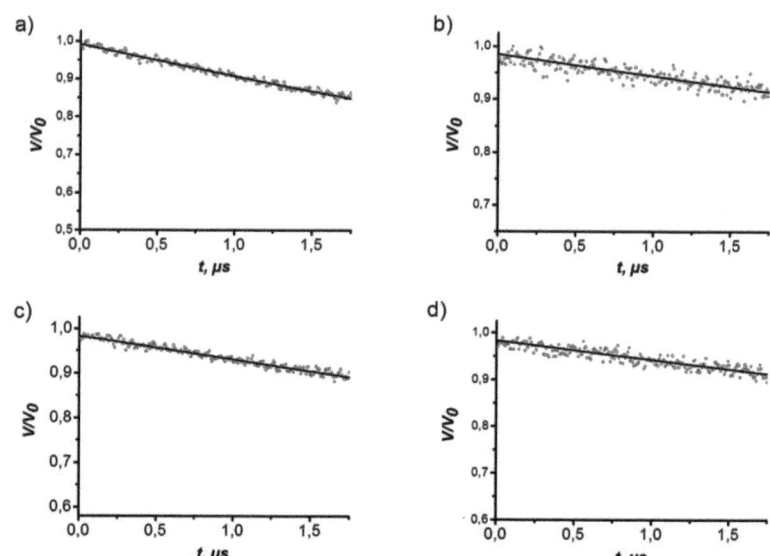

Figure A8. Dipolar evolution curves for single spin labeled LongHT oligonucleotides. (a) LongHT^{side} in Na$^+$-containing buffer, (b) LongHT^{mid} in Na$^+$-containing buffer, (c) LongHT^{side} in K$^+$-containing buffer, and (d) LongHT^{mid} in K$^+$-containing buffer.

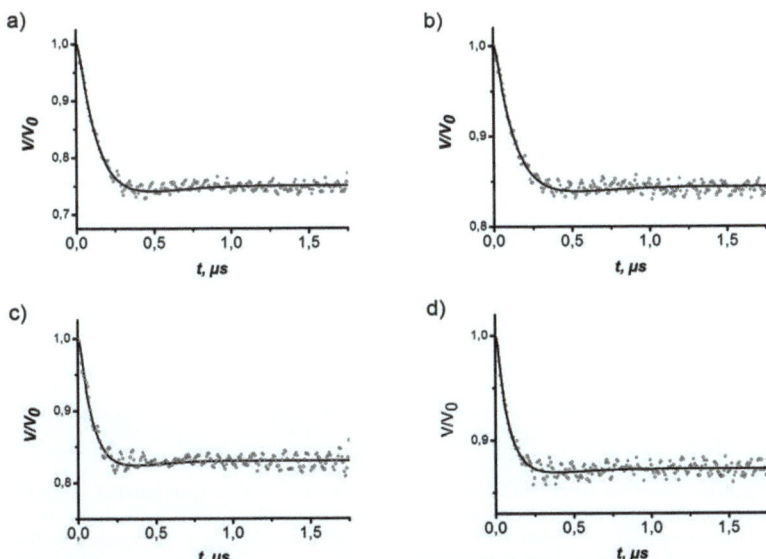

Figure A9. Dipolar evolution curves for single spin labeled LongHT oligonucleotides fitted with Tikhonov regularization. (a) LongHT^{side} in Na$^+$-containing buffer, (b) LongHT^{mid} in Na$^+$-containing buffer, (c) LongHT^{side} in K$^+$-containing buffer, and (d) LongHT^{mid} in K$^+$-containing buffer.

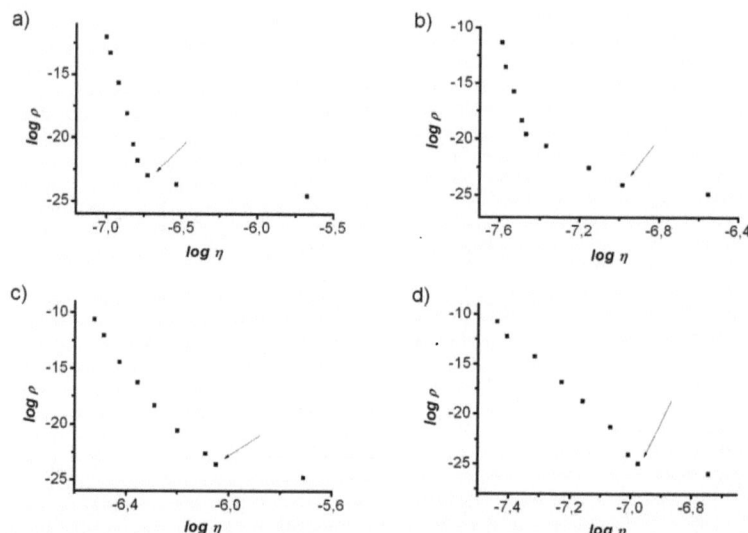

Figure A10. **L-curves from Tikhonov regularization for double spin labeled LongHT oligonucleotides.** (a) LongHT[side] in Na$^+$-containing buffer, (b) LongHT[mid] in Na$^+$-containing buffer, (c) LongHT[side] in K$^+$-containing buffer, and (d) LongHT[mid] in K$^+$-containing buffer. Optimal regularization parameter is marked with an arrow.

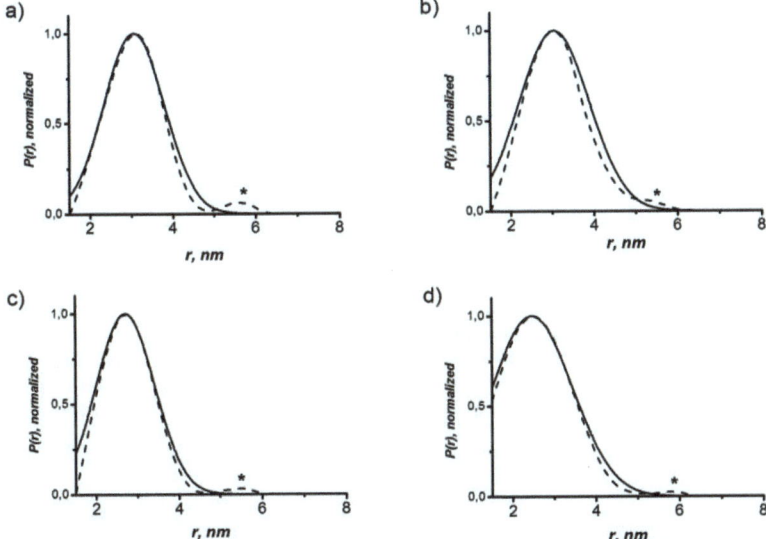

Figure A11. Comparison of distance distributions obtained from Tikhonov regularization (dotted line) with Gaussian-like ones (solid line). (a) LongHT^{side} in Na$^+$-containing buffer, (b) LongHT^{mid} in Na$^+$-containing buffer, (c) LongHT^{side} in K$^+$-containing buffer, and (d) LongHT^{mid} in K$^+$-containing buffer. Asterisks mark artefacts due to Tikhonov regularization.

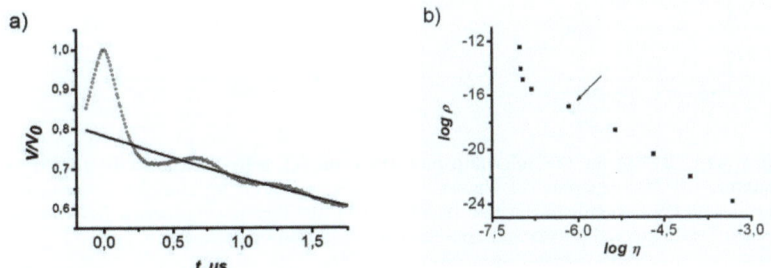

Figure A12. DEER on the DNA model helix in buffer solution. (a) The experimental dipolar evolution curve prior to background correction (circles) with the background function (solid line) for the double spin labeled DNA model helix. (b) The L-curve, the arrow indicates optimal regularization parameter α.

137

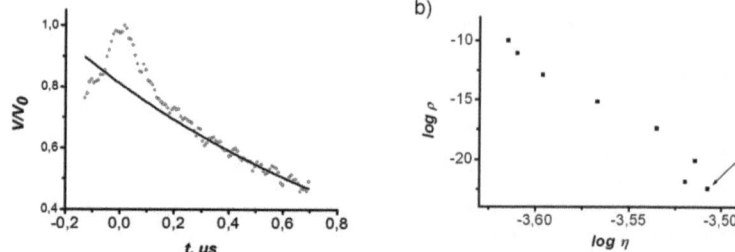

Figure A13. **DEER on the microinjected DNA model helix measured in the MS3 resonator.** (a) The experimental dipolar evolution curve prior to background correction (circles) with the background function (solid line) for the double spin labeled DNA model helix. (b) The L-curve, the arrow indicates optimal regularization parameter α.

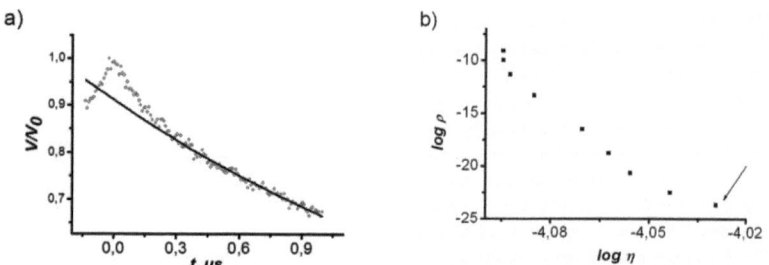

Figure A14. **DEER on the microinjected DNA model helix measured in the MD4 resonator.** (a) The experimental dipolar evolution curve prior to background correction (circles) with the background function (solid line) for the double spin labeled DNA model helix. (b) The L-curve, the arrow indicates optimal regularization parameter α.

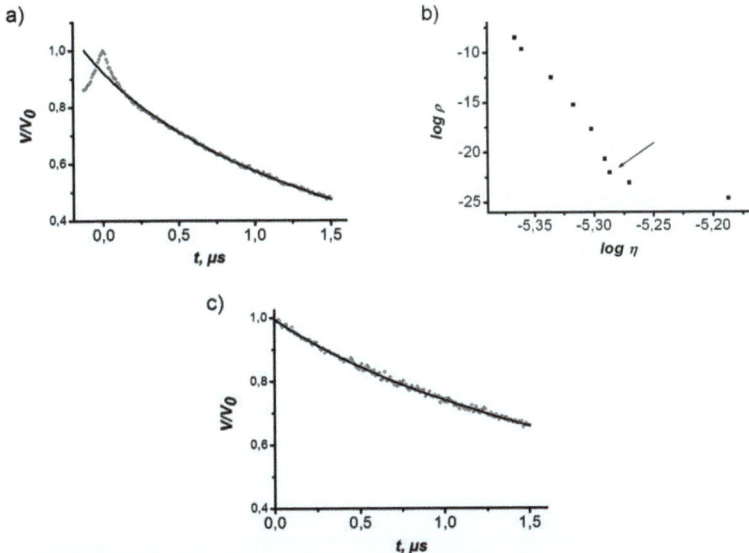

Figure A15. DEER on the microinjected HT-repeat. (a) The experimental dipolar evolution curve prior to background correction (circles) with the background function (solid line) for the double spin labeled oligonucleotide. (b) The L-curve, the arrow indicates optimal regularization parameter α. (c) The experimental dipolar evolution curve for the single spin labeled oligonucleotide with fit (solid line).

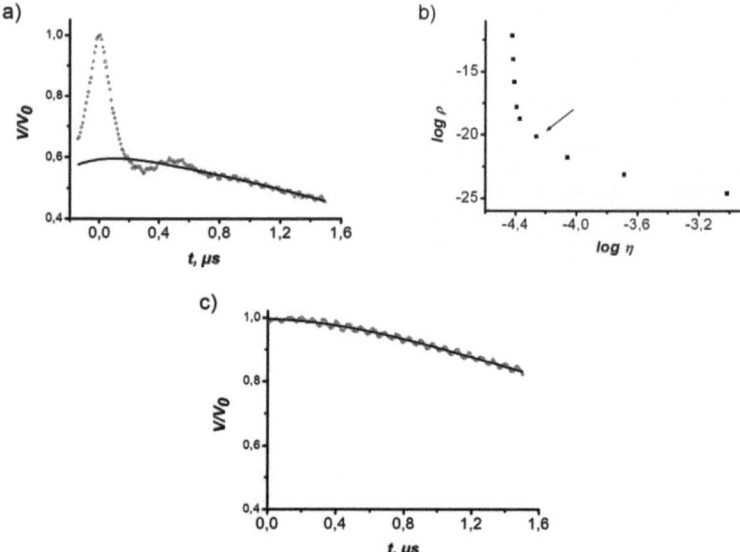

Figure A16. DEER on the HT-repeat in a salt-free buffer. (a) The experimental dipolar evolution curve prior to background correction (circles) with the background function (solid line) for the double spin labeled oligonucleotide. (b) The L-curve, the arrow indicates optimal regularization parameter α. (c) The experimental dipolar evolution curve for the single spin labeled oligonucleotide with fit (solid line).

i want morebooks!

Buy your books fast and straightforward online - at one of world's fastest growing online book stores! Environmentally sound due to Print-on-Demand technologies.

Buy your books online at
www.get-morebooks.com

Kaufen Sie Ihre Bücher schnell und unkompliziert online – auf einer der am schnellsten wachsenden Buchhandelsplattformen weltweit! Dank Print-On-Demand umwelt- und ressourcenschonend produziert.

Bücher schneller online kaufen
www.morebooks.de

VDM Verlagsservicegesellschaft mbH
Heinrich-Böcking-Str. 6-8　　　Telefon: +49 681 3720 174　　　info@vdm-vsg.de
D - 66121 Saarbrücken　　　　Telefax: +49 681 3720 1749　　　www.vdm-vsg.de

Printed by Books on Demand GmbH, Norderstedt / Germany